107
Advances in Biochemical Engineering/Biotechnology

Series Editor: T. Scheper

Advances in Biochemical Engineering/Biotechnology

Series Editor: T. Scheper

Recently Published and Forthcoming Volumes

Green Gene Technology
Research in an Area of Social Conflict
Volume Editors: Fiechter, A., Sautter, C.
Vol. 107, 2007

White Biotechnology
Volume Editors: Ulber, R., Sell, D.
Vol. 105, 2007

Analytics of Protein-DNA Interactions
Volume Editor: Seitz, H.
Vol. 104, 2007

Tissue Engineering II
Basics of Tissue Engineering and Tissue Applications
Volume Editors: Lee, K., Kaplan, D.
Vol. 103, 2007

Tissue Engineering I
Scaffold Systems for Tissue Engineering
Volume Editors: Lee, K., Kaplan, D.
Vol. 102, 2006

Cell Culture Engineering
Volume Editor: Hu, W.-S.
Vol. 101, 2006

Biotechnology for the Future
Volume Editor: Nielsen, J.
Vol. 100, 2005

Gene Therapy and Gene Delivery Systems
Volume Editors: Schaffer, D. V., Zhou, W.
Vol. 99, 2005

Sterile Filtration
Volume Editor: Jornitz, M. W.
Vol. 98, 2006

Marine Biotechnology II
Volume Editors: Le Gal, Y., Ulber, R.
Vol. 97, 2005

Marine Biotechnology I
Volume Editors: Le Gal, Y., Ulber, R.
Vol. 96, 2005

Microscopy Techniques
Volume Editor: Rietdorf, J.
Vol. 95, 2005

Regenerative Medicine II
Clinical and Preclinical Applications
Volume Editor: Yannas, I. V.
Vol. 94, 2005

Regenerative Medicine I
Theories, Models and Methods
Volume Editor: Yannas, I. V.
Vol. 93, 2005

Technology Transfer in Biotechnology
Volume Editor: Kragl, U.
Vol. 92, 2005

Recent Progress of Biochemical and Biomedical Engineering in Japan II
Volume Editor: Kobayashi, T.
Vol. 91, 2004

Recent Progress of Biochemical and Biomedical Engineering in Japan I
Volume Editor: Kobayashi, T.
Vol. 90, 2004

Physiological Stress Responses in Bioprocesses
Volume Editor: Enfors, S.-O.
Vol. 89, 2004

Molecular Biotechnology of Fungal β-Lactam Antibiotics and Related Peptide Synthetases
Volume Editor: Brakhage, A.
Vol. 88, 2004

Biomanufacturing
Volume Editor: Zhong, J.-J.
Vol. 87, 2004

Green Gene Technology

Research in an Area of Social Conflict

Volume Editors: Armin Fiechter · Christof Sautter

With contributions by

P. Aerni · N. Arrigo · H. Bauser · J. B. van Beilen
F. Bigler · H. Brunner · A. Einsele · F. Felber · C. Gessler
R. Guadagnuolo · F. Kessler · G. Kozlowski · M. Leisola
N. Malgras · M. P. Oeschger · A. Patocchi · M.-L. Plissonnier
Y. Poirier · J. Romeis · O. Sanvido · C. Sautter · T. Schlaich
A. Schrell · C. E. Silva · B. Urbaniak · P.-A. Vidi · F. Widmer

Springer

Advances in Biochemical Engineering/Biotechnology reviews actual trends in modern biotechnology. Its aim is to cover all aspects of this interdisciplinary technology where knowledge, methods and expertise are required for chemistry, biochemistry, micro-biology, genetics, chemical engineering and computer science. Special volumes are dedicated to selected topics which focus on new biotechnological products and new processes for their synthesis and purification. They give the state-of-the-art of a topic in a comprehensive way thus being a valuable source for the next 3–5 years. It also discusses new discoveries and applications. Special volumes are edited by well known guest editors who invite reputed authors for the review articles in their volumes.
In references *Advances in Biochemical Engineering/Biotechnology* is abbeviated *Adv Biochem Engin/Biotechnol* and is cited as a journal.

Springer WWW home page: springer.com
Visit the ABE content at springerlink.com

Library of Congress Control Number: 2007922570

ISSN 0724-6145
ISBN 978-3-540-71321-0 Springer Berlin Heidelberg New York
DOI 10.1007/978-3-540-71323-4

Springer is a part of Springer Science+Business Media

springer.com

Cover design: WMXDesign GmbH, Heidelberg
Typesetting and Production: LE-TEX Jelonek, Schmidt & Vöckler GbR, Leipzig

Printed on acid-free paper 02/3180 YL – 5 4 3 2 1 0

Advances in Biochemical Engineering/Biotechnology Also Available Electronically

For all customers who have a standing order to Advances in Biochemical Engineering/Biotechnology, we offer the electronic version via SpringerLink free of charge. Please contact your librarian who can receive a password or free access to the full articles by registering at:

springerlink.com

If you do not have a subscription, you can still view the tables of contents of the volumes and the abstract of each article by going to the SpringerLink Homepage, clicking on "Browse by Online Libraries", then "Chemical Sciences", and finally choose Advances in Biochemical Engineering/Biotechnology.

You will find information about the

- Editorial Board
- Aims and Scope
- Instructions for Authors
- Sample Contribution

at springer.com using the search function.

Attention all Users of the "Springer Handbook of Enzymes"

Information on this handbook can be found on the internet at springeronline.com

A complete list of all enzyme entries either as an alphabetical Name Index or as the EC-Number Index is available at the above mentioned URL. You can download and print them free of charge.

A complete list of all synonyms (more than 25,000 entries) used for the enzymes is available in print form (ISBN 3-540-41830-X).

Save 15%

We recommend a standing order for the series to ensure you automatically receive all volumes and all supplements and save 15% on the list price.

Preface

Green gene technology (GGT), understood as a part of modern biotechnology, has been on a steady, triumphal progression over the last ten years (ISAAA 2007, see the contribution by Einsele in this issue). This volume, jointly edited by Prof. Fiechter and me, deals with some actual scientific and socio-economic aspects with regard to genetically modified plants (GMP). Worldwide more than 100 million hectares of agronomical land are covered by GMP. This includes some prominent industrialised Western countries like the USA and Canada, a series of threshold countries like Argentina, Brazil, India and China, and a number of developing countries. Clearly, some of these countries have to deal with crop plant production and human nutrition in a very pragmatic way since, for example, India has to feed about a 1/5 of the world population on about 3% of the arable land. In contrast, the situation in Europe appears very different. Food supply is more than sufficient and comparably inexpensive. This surplus of food is on one hand convenient, since starvation has been largely unknown in Europe for about 50 years, with only comparatively few exceptions of socially peripheral individuals. On the other hand it makes the population careless about the future food supply. Even beyond mere food supply, Europe gained its cultural values from its agricultural success over the centuries. A single farmer became able to feed more and more people making them free to work outside of agriculture as a craftsman, artist, poet, scientist, engineer, mayor, administrative official, priest, philosopher, or soldier – to give only a few examples. In the public perception this connection between agronomy and cultural welfare is not sufficiently appreciated in Europe. Switzerland, geographically in the centre of Europe (although not a member of the political union) has the same cultural tradition, only somewhat shifted towards the more conservative mood common to mountain populations. In summary, a majority of Europeans, and the Swiss population in particular, are reluctant to new methods in agronomy.

Switzerland is probably the only country worldwide that has a moratorium on the commercial growth of genetically modified plants in its constitution. In contrast, the moratorium for GMP in the European Union between 1999 and 2004 was not legally binding. In Switzerland it was the population itself that established this moratorium into fundamental law by means of a referendum. Moreover, all Swiss legislation about gene technology, the so-called "Genlex", is probably the strongest law in place that attempts to prevent the abuse of gene technology worldwide. This includes, for example, protecting the dignity of organisms. We are not aware of any other country in the world that has extended the term dignity of organisms to plants at the level of making it law or that has included this extension in its release ordinance, which also regulates field experiments with GMP. Dignity of plants is particularly difficult to determine, since most of the categories known from dignity of animals, like natural behaviour or sexual propagation, are not applicable to crop plants, which have been bred to exhibit very unnatural behaviour. Potatoes, for example, are mostly pollen sterile, often seed sterile, and have been artificially selected for loss of their alkaloids in the tuber, which makes them an easy victim to many predators or pathogens. This exposure to its enemies would be a clear contradiction to animal dignity. However since we have little imagination about a plant's "well being", even ethical experts publicly convey a somewhat helpless impression with this issue.

This particularly strong position of the gene technology legislation, guided by the public mood against gene technology is remarkable in Switzerland, since this country owes a considerable part of its wealth to the chemical and pharmaceutical industries, which depend largely on biotechnology in their modern development. Industry research and developments dealing with GGT has consequently moved out. The research at the famous industry-owned Friedrich Miescher Institute in Basel is no longer engaged in plant research and the large rice genome project of Syngenta in Stein was first moved to England and then to the US. High regulation hurdles for a small country make it very unattractive to invest in deregulation for an agronomic area that is too small to get back the investment by selling seeds. With less than 100 000 ha, the largest crop area in Switzerland is maize, of which only a small proportion could be GM maize. Only a non-profit institution would be able to deregulate a GM crop plant. But the only biotechnology group at a federal research station that could have brought a GM line to market was closed down in 2005.

In contrast to this barren land with regard to the application of GGT, more than 80 basic research projects with GGT are ongoing in Swiss public research institutions, the universities and the federal research institutes (Farinata-Kramer 2005, http://www.forschung-leben.ch/download/BioFokus70.pdf). This is a remarkable number for such a small country. Swiss plant scientists are prominent authors in top-ranking international research journals. The projects range from very basic research like chromatin structure and function to fields with an apparent application perspective like disease resistance in crop plants. A small-scale field test should always be made as the last step for proof of concept at the end of such basic research projects with application perspective. Field tests in Switzerland are officially possible in spite of the moratorium, which concerns only commercial application. However the hurdle to get a permit is very high. There have only ever been three field tests with GMP in Switzerland and only one since 1992. It took an unaffordable 4 years to get permission for this harmless test performed in 2004, and financial expenses went beyond any relation to the scientific project costs. This money had to be spent on scientifically dispensable safety measures, attorney fees to support appeals in court, for professional guards and so on. Public research can not afford this time and expense a second time (see: Schlaich et al., in this issue). As a consequence, colleagues tend to do field tests in collaboration with colleagues abroad. The same experiment for which researchers in Switzerland were required to wait 4 years, submit 500 pages of applications and legal papers, and answer additional requests before permission was granted (in addition to the cost of all this), required US researchers to fill in a three-page form and agree to six weeks of evaluation by the authorities. This is apparently a very imbalanced situation for competition in research. As long as research stays in the lab, i.e. as long as it has no consequences, it is welcome in Europe. However, as soon as any application perspective becomes apparent, the resistance is extremely high because the final step for proof of concept needs an outdoors experiment. The legal situation, administrative officials, NGOs and the public mood collaborate very efficiently against research.

The huge mental discrepancy in society between research and application highlights that Switzerland in this sense is part of the European culture. Moreover by its size and the vehemence in the arguments of the opponent combatants in the public debate, Switzerland might even be a small core model for what happens in Europe on a larger scale. Therefore, when Prof. Fiechter asked me to join him in editing an issue of *Advances in Biochemical Engineering and Biotechnology* about green biotechnology, we immediately had the idea to focus on the Swiss situation: promotion must start at the centre of resistance otherwise it will be difficult to move anything. This is probably also true for changing the public mood on GGT. Promotion is necessary from the viewpoint of science, not in the sense that scientists should make political decisions – this is the field of the sovereign – but in the sense of insisting on dissemination of their knowledge and their rational conclusions also against a public majority. In contrast to industry, which has to sell products and thus has to please their customers, scientists working in public research institutions are not useful to society if they *only* prove experimentally what the public believes anyway. Who else if not public scientists should be the advocate of nature? GGT has huge potential for sustainable agriculture, for example, by reducing our dependence on agrochemicals and thus helping to preserve the environment, which shows that research in this area is more than justified.

How necessary scientists are who publicly communicate about the benefits and risks of GGT can be measured by how the public opinion is influenced by media in collaboration with a variety of NGOs, such as consumers, organic farmers and some groups that use concern about the environment for their own promotion. Usually, industry in the context of GGT is presented as a thoughtless, profit-hunting business – not considering that only industry has the capacity to develop a product from a scientific idea or a prototype to a useful and reliable product and bring it to market. Making profit with a product is not only permitted in our society, but also a driving force and a control instrument. Without profit, no expenses can be paid for the development and stakeholders would move their money away. The public on the other hand hardly recognizes that NGOs are as well enterprises, just hunting for members and donations. Often, the struggle for life leads such NGOs to argue against their own basic environment-protection ideas. In spite of this paradoxical situation there is a lot of public trust in these NGOs, which under these circumstances is not justified. Frequently, the media publish press releases from NGOs without even mentioning this fact, thus giving the impression that it is an editorial contribution. The Swiss TV Program SR1 has broadcasted Greenpeace's own video spots in the official evening news several times without designating these spots as Greenpeace-made. And whenever a contribution makes the impression to be too positive about GGT – even if it is fact-based – a second contribution must be broadcast that is sceptical of GGT, although this might just be factless scare mongering. "Well balanced" is the political term for this kind of misinformation. How should a non-expert TV watcher recognize the difference? Hence, the area of public information is dominated by a coalition of consumer protection agencies, organic and small-scale farmers' unions and environmental protection groups, which follow their own interest with mostly non-scientifically reasonable argumentation. The public and the voters can hardly get to an independent opinion about GGT under these conditions.

A small group of scientists realized some years ago that research should be engaged in public education and that more locally produced results are required for trust building with the public. These scientists asked the Swiss National Research Foundation SNF to establish a national research program about the benefits and risks of genetically modified plants. One of the ideas behind that project was to collect biosafety and benefit-research data from Switzerland in order to be able to argue with results from inside the country and to distribute the knowledge among the stakeholders of the GGT debate. Since then, this program (NFP59) has been granted and the project applications are under review. Although the program was not designed for this purpose, the hope of the politicians is that the program will deliver arguments about the moratorium in two years. At that time it will be discussed whether the moratorium will expire or if it will be prolonged. It is obviously convenient for the current government not to be under pressure for a decision about a coexistence regulation between GMP application and conventional or organic farming.

This political background and the public perception primarily supports organic farming with its roughly 10% of the agronomic production in Switzerland. This would be fine if this public debate would not at the same time discourage young people to engage themselves in the area of modern methods in agriculture and explore their putative benefits. Over the years, we have had fewer and fewer agro-biotechnology students since young people look not only for interesting fields for their studies, they also search for a topic which provides a perspective for their life to work on in the future. Switzerland has experience with moratoria and education: the moratorium in Swiss nuclear power plant production led to a draining of experts in this field in recent decades. It has to be assumed that the same will happen with GGT experts in the near future. Due to the small number of agro-biotechnology students, this topic has completely disappeared from the lectures on offer at ETH Zurich. In the view of the putative contribution of gene technology for sustainable agriculture, a lack of experts in the field is threatening the economic development of the whole country.

The application potential of GMP is broad. Up to now only herbicide tolerance and insect resistance have been the bulk traits. Their contribution to sustainability is already considerable, although these GMPs have not been designed for this purpose (Nillesen et al. 2005, see also Sanvido et al. in this volume). Currently the first crop plants with improved nutritional qualities like pro-vitamin A improved Golden Rice (a Swiss development) are under safety check for deregulation in several countries. Iron content is the next step in nutritional quality improvement. These nutritional traits are important for sustainability in the Western world but absolutely vital for developing countries. More complicated but under intensive study are traits for pathogen resistance, drought tolerance, and post-harvesting decay. A potato resistant to late blight (caused by *Phythophtora infestans*) is under development and could reduce the use of fungicides. Wheat resistant to *Fusarium* head blight would reduce the myco-toxin content of flour. In addition, the discussion of higher energy prices makes the production of renewable energy by GMP attractive again. Pharmaceuticals like antibodies or vaccines could be produced in GMP relatively inexpensively and without any risk of accompanying infections with human diseases. More putative applications will come up in the future. To miss all of these developments is a risk in its own.

The present volume of *Advances in Biochemical Engineering and Biotechnology* presents some of the few research topics that are currently under study in Switzerland. The socio-economic studies in this volume cover public perception, patenting, ethics, a comparison with the US, and the economy. Science contributions deal with fungal resistance (including field testing under Swiss conditions), biopolymer production, plastids and their compartments as target location for foreign products, biosafety with regard to out-crossing into wild relatives, putative impact of GMP to soil microflora and ecological impact of GMP over the last ten years of application.

In order to complete the picture, we have to admit that for various reasons many colleagues could not participate in this volume. The work and reviews of those groups can easily be found in the literature. Examples of topics are membrane ion transport, wheat genomics, transcription of plastids, transcriptional silencing, starch structure and biogenesis, apomixis, cell cycle regulation, genomic imprinting. Examples of applications with a focus on developing countries include the nutritional bio-fortification of rice and cassava of which Golden Rice and virus resistant cassava are the most advanced projects.

This volume should provide an idea of what is going on in Swiss GMP research and give an impression of the social and political environment in which this happens. Hopefully, it will create some understanding outside of Switzerland for the GMP research situation, their application in this country, and stimulate some readers to actively engage themselves in this research or its public communication. I thank Prof. Armin Fiechter for the opportunity to co-edit this special volume with him. It was his wish to publish this volume on this timely and controversial topic.

Zurich, March 2007 Christof Sautter

References

Nillesen E, Sara S, Wesseler J (2005) Do environmental impacts differ for Bt, Ht and conventional corn with respect to pesticide use in Europe? An empirical assessment using the Environmental Impact Quotient. IOBC/WPRS Bulletin 29(5):109–118

Contents

Adv Biochem Engin/Biotechnol (2007) 107: 1–11
DOI 10.1007/10_2007_055

Published online: 10 March 2007

The Gap between Science and Perception: The Case of Plant Biotechnology in Europe

Arthur Einsele

Internutrition, Postfach, 8035 Zurich, Switzerland
arthur.einsele@internutrition.ch

Abstract Although the global area of biotech crops continues to climb for the tenth consecutive year at a sustainable double-digit growth rate, the acceptance of biotech products from agriculture in Europe is still low. There is a gap between science and perception. It is a strong belief that the public turning against science and against GM food has been encouraged by the negative activities of NGO groups. Scientists have to overcome the purely risk-based discussion, and the benefits of plant biotechnology have to be made literally visible. GM food should be available, the benefits should be tangible and the consumer should have fun with such novel food. The gap could be reduced if genetically modified plants and the products thereof were regulated in the same way as classical products.

Keywords Benefits of plant biotechnology · Regulation ·
Perception of plant biotechnology · Global status of biotech crops

1 Facts

1.1 Global Status of Commercially Grown Biotech Crops

According to a study[1] the global area of biotech crops continued to grow for the tenth consecutive year at a double-digit growth-rate. The estimated global area of biotech crops for 2006 exceeded 100 million hectares; for the first time, the number of farmers growing biotech crops exceeded 10 million farmers in 22 countries. This represents an increase of 13% over 2005. Remarkably, the global biotech crop area increased more than fifty-fold in the first decade of commercialisation (see Fig. 1).

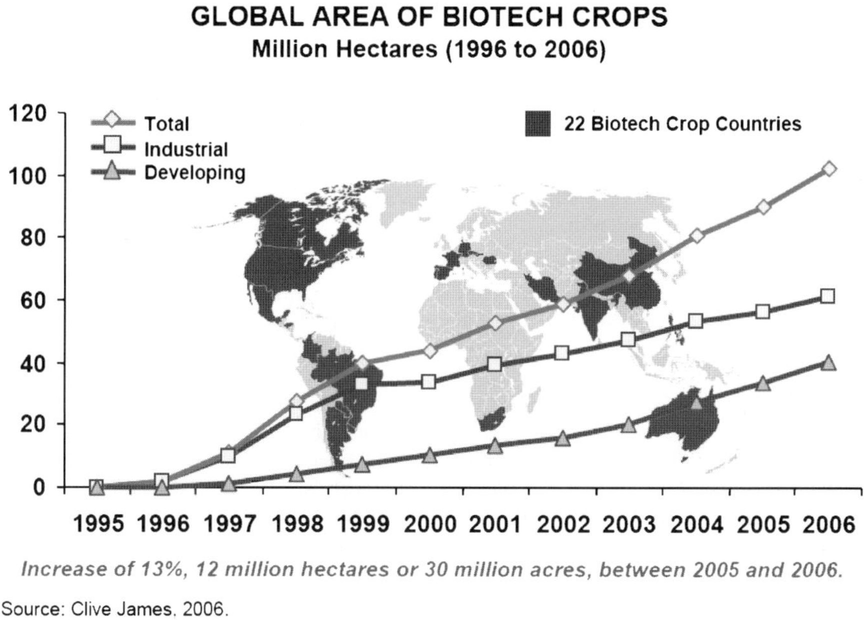

Fig. 1 Global area of biotech crops

The principal countries that grew biotech crops in 2006 included the USA, Argentina, Brazil, Canada, India and China; they represent 55% of the global biotech area.

However, there is also an increasing interest in Europe. Spain increased its area of Bt Maize in 2006 to 60 000. The collective Bt maize hectarage in the other five countries (France, Czech Republic, Portugal, Germany and Slovakia) increased over 5-fold from approximately 1500 hectares in 2005 to

[1] Global Status of commercialized Biotech/GM Crops: 2006; ISAAA Report No. 35

approximately 8500 hectares. In Europe there will be a trend towards an increase since the more agriculturally based countries of Eastern Europe which have recently joined the EU and those expected to join in 2007 and beyond have more small farmers who will apply biotech crops to meet their food/feed crop requirements. More, there are new signs for progress in European Union in 2004 with the EU Commission approving, for import, two events in biotech maize for food and feed use, thus signalling the end of the 1998 moratorium. The use of new crops, in conjunction with practical and equitable co-existence policies, opens up new opportunities for EU member countries to benefit from the commercialisation of biotech maize, which Spain has successfully deployed since 1998.

1.2 Benefits

The big benefit for all farmers is the sheer convenience of managing the crop: less spraying, less day-to-day monitoring of the crop and fewer tractor trips across the fields. This translates into real economic benefits in many cases.

Weed management is one of the main attractions for farmers in the use of herbicide-tolerant crops. In 1995, 86% of the US soya crop was treated with at least two different herbicides, and 23% used four or more. Control of weeds was good, but in 1994 it was still estimated that 7% of yield was lost, and that complex spraying regimes were needed to achieve this.

The availability of soya beans resistant to the broad-spectrum herbicide glyphosate gave many advantages, namely a much greater flexibility of application, leading to simpler treatment patterns and, in many cases, less spraying, less crop injury and no effect on follow-up crops because of low persistence in the soil [2].

Yield is a major concern for farmers growing any crop, and new technologies that enable them either to produce more or ensure a more consistent year-on-year yield are welcomed. In a recent report for the International Food Policy Research Institute, Mara et al. [3] compiled figures on the yield effects of various GM crops (Table 1).

Even in herbicide-tolerant oilseed rape in Canada, the small yield decrease actually still leads to increased profitability because of lower input costs. Therefore, input costs can be considerably reduced for some crops. Cotton, for example, is a notoriously difficult crop to manage since the bollworm can destroy much of the cotton boll from inside and therefore is difficult to treat effectively by spraying. In China, Pray et al. [4] reported that savings equivalent to up to $ 200/ha could be made in insecticide by planting Bt cotton.

Savings in the USA have also been large. It has been estimated that American cotton farmers benefited by $ 97 million in 1998 [5]. In the period 1996–1999, use of pesticides on cotton crops approximately halved, at the same time the yield per hectare increased by 7%.

Table 1 Yield effects on variable crops

Crop	Growing area	Average yield increase bushels/ha	Average net profit increase $/ha
Insect-resistant maize	US corn belt	26.7	148.4
Herbicide-tolerant oilseed rape	Canada	– 4.7	27.9
Herbicide-tolerant soya bean	North Carolina	6.7	43.6

The facts and figures for the overall impact will vary for particular crops with the growing area and the season. But, it is clear that the benefits for many crops are sufficient to account for the enthusiastic take-up of the technology by farmers in several countries.

A new study reports experiments in China concerning the commercialising of GM rice [6]. The paper studies two of the four GM varieties that are now in farm-level preproduction trials, the last step before commercialisation. Surveys of randomly selected farm households that are cultivating the insect-resistant GM rice varieties, without the aid of experimental station technicians, demonstrate that when compared with households cultivating non-GM rice, small and poor farm households benefit from adopting GM rice by both higher crop yields and reduced use of insecticides, which also contributes to improved health.

1.3 Regulatory Status

Before 1990, the biotechnology regulation and policies were similar in most countries, namely in West European countries and in the Americas. But, after 1990 the EU and its member states have moved towards stringent approval and other regulatory standards. The new rules in most European countries followed the precautionary principle. This means that the risk to public health or the environment is existent as long as scientific knowledge of the respective risk has not demonstrated the overall safety. With this development the regulation between the EU and the USA began to diverge. In contrast to Europe, US policy makers have embraced agricultural biotechnology. The new products were seen as a new and innovative way to produce food and the new method is accepted as having no other influence on the health and the environment than the traditional methods.

Following the precautionary principle, in the EU only few agricultural biotech products have been approved for commercialisation. The fact that the regulation has diverged between the main regions can be called regulatory polarization [7]: an increasing gap is developing between agrobiotechnology-promoting countries (Argentina, Canada and USA) and the agrobiotechnology-restricting countries (European Union). Regulation in the Americas is focussed on the products and is mainly science-based. In the EU the regulation is process-based. This means that as soon as a product is based on a process, which includes genetic engineering, it will be regulated differently and much more profoundly. It is a fact that the stricter regulation in Europe has added nothing to the safety of agrobiotech products and the regulation has not improved the trust in biotech products. On the contrary, as soon as a product fails to fulfil all regulatory requirements, although it is scientifically proven safe, this product is seen as very risky and even dangerous. Bernauer [7] has found the following text:

"*In the US products are safe until proven risky*
In France products are risky until proven safe
In the UK products are risky even when proven safe
In Switzerland products are risky especially after they have proven safe!"

2 Perception

This leads directly to the problems of perception. It can be seen that the public perception of agrobiotech products is very different from the view based on scientific facts. Therefore, we call this a huge gap between science and perception. Public attitudes are studied with opinion polls; in Europe there is the so-called Eurobarometer, which is published frequently. There are many of these kind of polls; some of them will be discussed and interpreted in this section.

2.1 Opinion Polls

In a recent study [8] it has been reported that Americans are largely unaware of GM food, both of its presence in their lives and of its wide application in food production. In addition, most Americans have little understanding of the general facts of genetic engineering. They are unfamiliar with the laws and safety testing regarding GM food but are generally familiar with which agencies are responsible for such overseeing. This is very important, since Americans believe that the agencies are doing a good job and that the approved products are safe and reliable. It is quite different in Europe, where there is no trust in the agencies involved.

Americans report interest in a variety of topics related to GM food, and say they would watch television shows about the topic, though most report that they have never actively sought information about agricultural biotechnology. Most respondents said they would search the Internet if looking for information about GM food. Americans opinion towards GM food remains uncrystallised and uninformed. While Americans say they are interested in the topic, they have not yet been stimulated enough to actively seek information about the technology, and have had little passive exposure to the topic.

In the United Kingdom, a survey was carried out in 2001. About 1000 respondents were asked to say what benefits or risks were associated with GM crops.

Interestingly, 22% said there were no potential benefits, with 31% don't know; for potential risks, 13% said none, and 38% didn't know. Just over half of the sample was therefore unable to identify either risks or benefits without prompting.

Equally illuminating is the fact that, of those identifying potential risks (49%) half said, "risks to health". This was taken to reflect the personal concerns about food safety in the UK following the BSE crisis, a number of high-profile food poisoning incidents and general lack of trust in government to protect them.

Another survey prepared by the Food Standard Agency [9] concluded in 2003 that most consumers do not have entrenched views on GM food, but there is a suspicion of GM, and there is a lack of readily understood information. Although research has shown that concerns about GM food has decreased over the past 3 years (consistent with the Eurobarometer) it appears that for many people any consumer benefits from GM food remain unclear and unproven.

The potential impact of GM crops on the environment was the issue that gave rise to most concern and emerged in all the activities undertaken by the Food Standard Agency. The safety of GM food was less of an issue, but suspicion and concern still surround the subject.

A particular worry was that once GM crops were released into the environment, there could be no turning back and that, in turn, could restrict choice between GM and non-GM food through cross-contamination.

2.2 Eurobarometer

Eurobarometer surveys on biotechnology were conducted in 1991, 1993, 1996, 1999 and in 2002. After a decade of continuously declining optimism in biotechnology the trends reversed in 2002 [10]. An index of optimism shows an appreciable change from the declining trend of the years 1991–1999. In the period 1999–2002, optimism has increased to the level seen in early 1992. The rise in optimism holds for all EU member states with the exception of Germany and the Netherlands, where such a rise was observed between 1996 and 1999.

A majority of Europeans do not support GM foods. They are judged not to be useful and to be risky for society. This is also true for GM crops since they are judged as moderately useful but as risky as GM food. The greater opposition to GM food over GM crops, reflected in perceptions of lower usefulness, higher risk and lower moral acceptability, suggests that Europeans may be more concerned about food safety than the environmental impacts of agrofood biotechnologies. This exactly reflects the gap between the scientific facts and public perception. This is because the facts show that neither GM crops nor GM food have ever been shown to be more risky than traditionally bred crops and products made thereof. As a consequence of this negative perception for GM crops, most countries in the EU have a *de facto* moratorium on the commercial exploitation of GM crops. A better support for GM food is seen in only four countries, namely Spain, Portugal, Ireland and Finland.

The Eurobarometer shows that the Europeans have less confidence in their governments than the citizens in the USA. Whereas around 70% of Europeans have confidence in doctors, university scientists, consumer organisations and patients' organisations, less than 50% have confidence in their own government and in industry.

There are mixed opinions on the acceptability of buying and consuming GM food (see Table 2). The most persuasive reason for buying GM food is the health benefit of lower pesticide residues, closely followed by an environmental benefit. The price was not very much the decision supporting point; price is apparently the least incentive for buying GM foods. However, what people say and what they do in the shop are sometimes rather different. Since

Table 2 Hypothetical purchasing intentions of European consumers. This shows the answers to the question: Do you agree/disagree to purchase and to eat a GM food if .... These data from the Eurobarometer are similar to those reported in an overview made by the Food Standard Agency, UK [9]

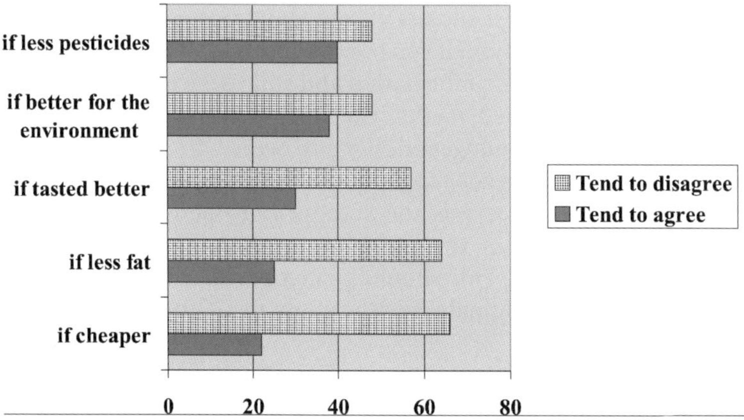

the question about buying GM food is mostly a "theoretical" one, it is likely that people are thinking as a citizen rather than as a consumer. Since there is no GM food currently in the shops, the respondents are not able to reflect on actual or related experiences. With questions about GM food the issue is genuinely novel.

For all the hypothetical situations, there are more Europeans saying they would not buy or eat GM food than those saying they would.

2.3 The Role of NGOs

It is a strong belief that the public turning against science and in particular against GMO food has been encouraged by negative articles in newspapers and the activities of lobby groups against plant biotechnology, namely some NGO groups.

It is clear that the current public mode is broadly anti-science. Greenpeace and other organisations regard GM crops as so immoral or dangerous that they break the law, invade fields and destroy farmer's property. They want to stop all experiments, which are designed to establish what positive impacts GM crops will have for the environment. But, by their actions they imply that they already know the outcome, namely that GM crops are dangerous.

The most aggressive and visible opponents of GM crops and GM food are Greenpeace and Friends of the Earth. They both have huge memberships and influence and they have decisively moved into the anti-science camp. They often argue that evidence from company scientists must be discounted because companies have to make profits. Indeed, one has to always look carefully at company claims (as for any other claim) but this does not necessarily means that they are wrong or those researchers who work for a biotech company have no concern for humankind. However, we should examine equally sceptically claims made by Greenpeace and Friends of the Earth.

A recent example is the story about "golden rice": Greenpeace seems to have developed an allergy against success stories about GM crops.

It is hardly necessary to mention, if you see the latest Greenpeace hoax [11] on golden rice, that it is full of misinformation and still sticks to the old myths and produces some new falsified myths. Greenpeace seems to have a real problem with success stories about genetically engineered crops: it is a classic piece of pseudoscience – the Greenpeace spin-doctors became nervous.

Another target of Greenpeace was the success story about Bt cotton in China, where again they did not shy away from presenting pseudoscience (see their original report [12]). This should be compared with the response of the Chinese author of the original scientific study that Greenpeace grossly distorted [13].

It is hard to understand why an organization like Greenpeace can still maintain its charity status. It is hard to see any charity and mercy at all in their stance against GM crops. In the face of 250 000 to 500 000 children going blind every year we can see no understanding for such a purist point of view, shutting out a promising solution for the dramatic nutritional problems. We simply cannot afford in the face of a humanitarian catastrophe of major proportions such a painstaking methodological debate with half-truths and blunt lies shutting out one particular solution even before it can become reality.

But, it is a fact that those NGOs are getting more and more influence and are politically powerful, especially in Europe, with political Green parties and some Socialist parties. Since they are perceived as trustful, this influence is a fact and scientists have to worry about it.

An interesting view has been published recently by Dick Taverne [14], who supports the promotion of an evidence-based approach to scientific issues. He confirms that the past few decades have witnessed a growing influence of "green" activists who approach environmental issues with a semi-religious zeal and seemingly little regard for evidence. The increasing prominence of these viewpoints in the media and in the political arena is a significant cause for concern. It questions the future of the plant biotech enterprise.

3 Factors Affecting the Gap Reduction

3.1 The Risk–Benefit Imbalance

In a first step it has to be realized that the benefits of GM plants and the products thereof have to be understood by the public at large. Simultaneously, people have to realise that the dire consequences suggested by some opponents have not happened. Consumers want information about GM food that is factual, user friendly and relevant to their lives. This information must have an emotional element.

Scientists have to overcome the purely risk-based discussion, which follows the line of the precautionary principle. Today's generation belongs to a population that is very risk-oriented; society is in a rather wealthy situation in Europe where any risk can influence wealth and therefore has to be avoided. As long as GM technology is looked as being mainly risky, this technology will not be preferred.

The promotion of risks, as trumpeted by opponents, has to be dismantled and the speaker for these risks has to be clearly discredited.

Finally, the benefits have to be made literally visible. This means that GM food has to be made available in Europe. It is a vicious circle since the opponents claim that the population does not want GM food and as a consequence farmers and retail stores do not offer this kind of food, again based on the assumption that consumers do not want it. In reality, up to 20% of the consumers in Switzerland [15] would be ready to buy food based on GM plants if available. This is almost double the population that is dedicated to buying organic food. As long as we have no GM food on the shelves we do not have a real choice for the consumer. GM food should be available, the benefits should be tangible, and the consumer should have fun with such novel food.

3.2 The Regulatory Dilemma

We are on a slippery slope: opponents and proponents ask for efforts to establish ever stricter, more complex and costly regulations. This will be very difficult to implement and to divorce from scientific evidence for any risk. This may be asked from different standpoints. Opponents ask because they want a regulation to avoid any imaginary risk. Proponents think that with strict rules we will have sooner public acceptance. However, today the opposite is already true: the rules are so strict that they can hardly be fulfilled and if opponents find facts that are not regulated, then they ask for more regulations. This is a trap! Regulation is not the answer. Firstly, it does not help to improve acceptance and, secondly, evidence based on the experiences of the last 10–15 years has shown that GM food and the products thereof are "normal" products and do not need any special regulation.

Therefore, a driving force to reduce the gap between public perception and reality would be the redesign of the current regulatory system for plant biotechnology. GM plants and the products thereof should be regulated in the same way as "classical products".

4 Conclusions

The global area of biotech crops has continuously grown during the last 10 years at a double-digit growth rate and represents more than 90×10^6 ha today. More than 8.5 million farmers in 21 countries plant this area.

There are numerous reports about the benefits, mainly advantages for the farmers. These positive effects can be expressed in less spraying and less day-to-day monitoring of the crops, which results in a real economic benefit in most cases.

In contrast to the positive attitude of a science-based position of the public at large in the Americas, public perception in Europe is very different. There is a huge gap between the science-based facts of the safety and the benefits of biotech products compared to the negative perception almost all over Europe.

The strict regulatory and labelling rules have not changed the negative attitude in Europe. On the contrary, as soon as a product does not fulfil all regulatory requirements but it is scientifically proven safe, this product is seen as very risky and even dangerous.

It is a strong belief that the public turning against plant biotechnology has been encouraged by the negative reports in newspapers initiated by lobby groups against genetic engineering.

There is no easy turn-around proposal in Europe to get a better acceptance of biotechnology-derived food in Europe. Two driving forces could be of interest: first, a demonstration and a taste of the benefits to the public combined with a relaxed risk discussion based on science; in a second phase, society should consider a redesign of the regulatory principles for the approval of products from plant biotechnology. They should rather be regarded as "regular products".

Acknowledgements The author thanks Klaus Ammann, former director of the botanical garden in Berne, for the access to his huge database on science- and non-science-based argumentation.

References

1. James C (2004) ISAAA Briefs 35-2006: Global status of commercialized biotech/GM crops: 2006
2. Agricultural Biotechnology in Europe (2002) Economic impacts of crop biotechnology, Issue Paper 5, October 2002
3. Mara MC et al. (2002) The payoffs to agricultural biotechnology: an assessment of the evidence, EPTD discussion paper 87. International Food Policy Research Institute, Washington, DC
4. Pray et al. (2001) Impact of Bt cotton in China. World Dev 29:813–825
5. Frisvold et al. (2000) Adoption of Bt cotton; regional differences in producer costs and returns. Proc Beltwide Cotton Conf 2:237–240, Memphis TN, NCAA
6. Huang J et al. (2005) Insect-resistant GM rice in farmers' fields: assessing productivity and health effects in China. Science 308:688
7. Bernauer T (2003) Genes, trade and regulation: the seeds of conflict in food biotechnology. Princeton University Press, Princeton, NJ
8. Hallmann WK et al. (2004) Americans and GM food: knowledge, opinion and interest in 2004, FPI publication number RR-1104-007. Food Policy Institute, Rutgers, NJ
9. Food Standards Agency (2003) Consumer views of GM food
10. Gaskell G, Allum N, Stares S (2003) Europeans and biotechnology in 2002, Eurobarometer 58.0, 2nd edn, March 2003
11. Greenpeace (2005) Golden rice: All glitter, no gold. Greenpeace
12. Xue D (2002) A summary of research on the impacts of Bt cotton in China
13. Wu K (2002) A brief statement on the studies of the ecological impact of Bt cotton conducted by Dr. Kongming Wu's lab, Institute of Plant Protection, Chinese Academy of Agricultural Sciences, Beijing
14. Taverne D (2005) The new fundamentalism. Nat Biotechnol 23(4):415
15. Longchamp C (2003) Final report for "Gentech-Monitor". GfS-Research Institute, Berne

Invited by: Professor Fiechter

Adv Biochem Engin/Biotechnol (2007) 107: 13–39
DOI 10.1007/10_2007_049

Published online: 5 April 2007

Biotechnology Patenting Policy in the European Union – as Exemplified by the Development in Germany

Andreas Schrell[1] · Herbert Bauser[2] · Herwig Brunner[2] (✉)

[1]European Patent Attorney, Leitzstrasse 45, 70469 Stuttgart, Germany

[2]Fraunhofer Institute for Interfacial Engineering and Biotechnology, Nobelstrasse 12, 70569 Stuttgart, Germany
herwig.brunner@igb.fhg.de

Abstract Patenting of biotechnological inventions is an important concomitant side effect of progress in this field, but also a matter of dispute in the public. In this paper, the significance of and the prerequisites for patenting are reviewed, and the principal requirements for biotechnology patents in the signatory states of the European Patent Convention (EPC) are summarized. This is followed by a report on the historical development of biotech-patent legislation in Europe and in Germany as one contracting state to EPC and member state of the European Union. Characteristic features of the patenting policy in Europe and Germany are illustrated by critical examples of biotechnology patents or patent applications. Some examples illustrate the influence of the European Union's national states' case laws after these had crystallized into the EU Biotechnology Directive (1998), which later was adopted by the European Patent Organization into its Implementing Regulations (2001) and was implemented into national patent acts. Some frequent objections against patenting in modern biotechnology are considered. More and better information about prerequisites, consequences, and opportunities of patenting in biotechnology, if conveyed to science and technology scholars as multipliers, may help to rationalize public discussion.

Keywords Biotechnology patent regulations (Europe, Germany)

Abbreviations

BGBl	(German) Federal Law Gazette
BGH	(German) Federal Supreme Court
BGHZ	Decision of the (German) Federal Supreme Court in a Civil Case
BPatG	(German) Federal Patent Court
BPatGE	Decision of the (German) Federal Patent Court
EC	European Community (since 1993: European Union)
EPC	European Patent Convention
EPO	European Patent Office
EU	European Union
GPA	German Patent Act
GPTO	German Patent and Trademark Office
HGP	(German) Human Genome Project
JPO	Japan Patent Office
OD	Opposition Division of the European Patent Office
R&D	Research and Development
TRIPS	Agreement on Trade Related Aspects of Intellectual Property Rights
USPTO	United States Patent and Trademark Office

Abbreviations of patent journals:

GRUR	Gewerblicher Rechtsschutz und Urheberrecht (Protection of Industrial Rights and Copyright)
Mitt.	Mitteilungen der deutschen Patentanwälte (Communications of the German Patent Attorneys)
OJ EPO	Official Journal of the European Patent Office

1 Introduction

Since the mid-1970s, patenting in biotechnology became a public issue, whereas for the relevant applied research laboratories the patenting began a century before with a US patent granted in 1873 to Louis Pasteur for a "yeast, free from organic germs of disease, as an article of manufacture" [1]. It was the discussion about patenting a bacterium as a living being [2, 3] together with the invention of genetic engineering [4–6] in the early 1970s, which stirred the public interest and caused silent consent and loud protests in several European countries. In Germany and some other countries, a distrusting and even hostile attitude towards biotechnology and, in particular, against genetic engineering prevailed in the 1970s and 1980s [7,8]. In the meantime this passionate opposition has diminished and given room to a more rational approach, at least in several fields, the green gene technology still being excluded and dramatically hampered by legislation in Germany. The reservations about *patenting* of biotechnological inventions have additional roots, and the more important of those have to be examined.

On the other hand, a great part of the public eagerly expects the development of novel medical drugs on the basis of genetic engineering against cancer, cardiovascular and degenerative diseases, rheumatism, and sundry other health problems. Many pharmaceutical companies invest impressive amounts of Euros into those developments. Companies need patent protection for their novel drugs for the time scheduled. The same is true for biotechnological activities in other fields, such as in food, agricultural, and environmental industry. This situation poses an obligation on all parties to consider objectively the issue of defining and patenting inventions in the field of biotechnology, and to withstand prejudices and misleading slogans.

In this paper we want to assist the reader in recollecting the definition and prerequisites of patenting, and to guide him/her on a short trip through the historical development of patenting legislation in biotechnology. We restrict ourselves to the European patent policy and to Germany as an example for a European country. Patent regulations, such as the European Patent Convention (EPC) [9] and the German Patent Act [10], reflect both the technical and legal fundaments on which these regulations are based and are functioning. They may therefore be considered as being proven in practice. Furthermore, we have to keep in mind, that the patent protection always refers to *technical inventions* thus being valid also for the field of *bio*technology. This restriction to technical inventions proves the slogan "patents on life" to be irrelevant.

2
The Significance of Patents

The principles underlying the patent system are very simple: The state, as represented by the patent office, grants the inventor an exclusive monopoly on his invention for a limited time and for a defined region—i.e. the national territory—in return for his disclosure of the invention to the public, so that the public will be able to learn and use the invention within the framework of patentee's rights and, after expiry of the patent, without any of such limits. To express it in the words of Abraham Lincoln, "the patent system added the fuel of interest to the fire of genius".

Filing a patent application for an invention usually results in the invention being published either as application or as patent. The inventor, principally not being forced to make his invention publicly available, therefore adds to mankind's knowledge and capabilities. As a reward the inventor obtains the right to exclude others from his invention, guaranteeing him an advantage in the market or license fees in case third parties are allowed to use the invention. The significance of the patent system must be seen primarily in its rewarding and motivating effect for all those who contribute to mankind's capabilities.

Legal protection of inventions has been known for several hundreds of years, e.g. in Venice (1474), England (1624), USA (1790), France (1791), Austria (1794), Prussia (1815), and the Netherlands (1817). Royal and Imperial privileges for inventors (England since 1449, France since 1649, Holy Roman Empire since about 1531) preceded legislation yet developed and already exerted patenting criteria and procedures [11, 12]. Both this long history and the global persistence and pervasiveness provide ample support for the above consideration and the worldwide belief that a patent system based on the above principle of exchange is necessary for guaranteeing technical progress and development.

Another essential item to understand is that a patent primarily represents an exclusive right. It enables the patentee to exclude others from using his invention. The patent itself, however, does not unconditionally allow the patentee to use the invention. Whether a technology—be it a patented or non-patented technology—is allowed to be used depends (i) upon the general law system applicable in the state of question, e.g. on laws governing the protection of drugs, embryos, on laws relating to public health, on criminal law etc. And of course, it depends (ii) upon patents of other proprietors. Thus, a patent protects the realization of a novel technical concept against imitation. But it does *not* constitute a "real" right—i.e. a right *in rem*—to the subject matter of that patent.

3
Prerequisites for Patents

3.1
General Patentability Prerequisites for Inventions

For a given item of technology to be patentable, it must be of a kind, which is inherently eligible for patenting:

- it must be new,
- it must involve an inventive step,
- it must be susceptible of industrial application,

- it must be sufficiently disclosed to the skilled person and
- shall not be otherwise excluded by statutory exceptions to grant.

In these respects, the patent systems in the world are much alike, although the US-patent system still encompasses a few peculiarities such as the first-to-invent system, or the specific novelty regulations, e.g. the grace period for academic inventors, which is also granted in some other national patent laws [13, 14]. However, in particular with respect to patenting biotech inventions, a trilateral project between the European Patent Office (EPO), the Japanese Patent Office (JPO), and the US Patent and Trademark Office (USPTO) aims to harmonise the standards of certain patentability criteria, namely inventive step and sufficiency of disclosure [15]. The following considerations are based on the currently applicable law of the EPC, which closely resembles the stipulations of the German Patent Act.

3.2 Eligibility Requirements for Patenting under the EPC

The question of what is in principle patentable under the EPC has to be answered on the basis of Articles 52 and 53 EPC [9]:

Article 52: Patentable Inventions

1. European patents shall be granted for any inventions, in all fields of technology, provided that they are new, involve an inventive step and are susceptible of industrial application.
2. The following in particular shall not be regarded as inventions within the meaning of paragraph 1:
   (a) discoveries, scientific theories and mathematical methods;
   (b) aesthetic creations;
   (c) schemes, rules and methods for performing mental acts, playing games or doing business, and programs for computers;
   (d) presentations of information.

Although the law itself does not explicitly indicate what an invention is, and what not, case law both in Europe and Germany define an invention generally as a practical teaching which requires the claimed subject matter to have a technical character, which is capable of being realised repeatedly and provides a solution to a problem based on technical consideration. The term "problem" merely indicates that the skilled person is faced with some tasks; it does not mean that the problem is of a particular severe nature and poses great difficulty to the person aiming to solve the problem.

A teaching is of technical nature and therefore in principle patentable if it makes use of the operation of controllable natural forces other than the working of human intelligence to achieve a causally predictable result. In contrast to a discovery that contributes and adds solely to the knowledge of mankind, an invention usually contributes and adds to the capabilities of mankind to solve a problem, be it of technical or non-technical nature [16]. Thus, discoveries, scientific theories and mathematical methods as well as business methods or games are not patentable. Article 52 (1) EPC furthermore precisely clarifies that patents may only be granted if the invention is new, involves an inventive step and is applicable in industry.

Article 53: Exceptions to Patentability

European patents shall not be granted in respect of:
(a) inventions the commercial exploitation of which would be contrary to "ordre public" or morality; such exploitation shall not be deemed to be so contrary merely because it is prohibited by law or regulation in some or all of the Contracting States;
(b) plant or animal varieties or essentially biological processes for the production of plants or animals; this provision shall not apply to microbiological processes or the products thereof;
(c) methods for treatment of the human or animal body by surgery or therapy and diagnostic methods practiced on the human or animal body; this provision shall not apply to products, in particular substances or compositions, for use in any of these methods.

Article 53(a) EPC aims to exclude inventions from patent protection if they stand in unambiguous contradiction to the widely accepted ethical and social standards in our society. Accordingly, the patent law must be applied in a way that respects the fundamental principles of social and individual life guaranteeing the integrity and dignity of individuals. On the other hand patent law is not the watchman of morals. Patent Offices are technically oriented registration and administration authorities and not designed to correct or control the development of technology.

Article 53(b) EPC excludes plant and animal varieties from patent protection, i.e. a homogenous group of organisms within the lowest rank, i.e. within the species rank in the kingdom of plants and animals.

According to Article 53(c), neither therapeutic nor surgical treatments nor diagnostic methods practiced on the human or animal body are patentable under the EPC. Such procedures are not considered susceptible of industrial application. Therefore, unlike the USPTO, both the German Patent and Trademark Office (GPTO) and the EPO do not allow method-of-treatment claims. It should be understood, however, that formulating substances for use in such methods could be protected.

Both the German and the European Law, foresee further exceptions to the above-defined concept of patentability as is evident from the following.

3.3
The Conditions for Biotechnological Patents in the EPC Contracting States

With respect to biotechnological inventions, particular attention has to be drawn to the Implementing Regulations to the EPC [17]. In particular Rules 26 to 29 shed some light on the general stipulations to patentability as seen before, in particular Art. 52 (2) EPC and Art. 53 EPC. Rules 26 to 29 of the Implementing Regulations have been incorporated into the EPC in view of the EU-Directive on the legal protection of biotechnological inventions [18], as discussed in more detail below.

Rule 26: General and Definitions

(corresponds to Art. 2 EU Directive [18])

1. For European patent applications and patents concerning biotechnological inventions, the relevant provisions of the Convention shall be applied and interpreted in accordance with the provisions of this chapter. Directive 98/44/EC of 6th July 1998 on the legal protection of biotechnological inventions shall be used as a supplementary means of interpretation.
2. "**Biotechnological** inventions" are inventions, which concern a product consisting of or containing biological material or a process by means of which biological material is produced, processed or used.
3. "Biological material" means any material containing generic information and capable of reproducing itself or being reproduced in a biological system.
4. "Plant variety" means any plant grouping within a single botanical taxon of the lowest known rank, which grouping, irrespective of whether the conditions for the grant of a plant variety right are fully met, can be:
   (a) defined by the expression of the characteristics that results from a given genotype or combination of genotypes,
   (b) distinguished from any other plant grouping by the expression of at least one of the said characteristics, and
   (c) considered as a unit with regard to its suitability for being propagated unchanged.
5. A process for the production of plants or animals is essentially biological if it consists entirely of natural phenomena such as crossing or selection.
6. "Microbiological process" means any process involving or performed upon or resulting in microbiological material.

Rule 27: Patentable Biotechnological Inventions

(corresponds to Art. 3(2) and 4(2) EU Directive)

Biotechnological inventions shall also be patentable if they concern:

(a) biological material which is isolated from its natural environment or produced by means of a technical process even if it previously occurred in nature;

(b) plants or animals if the technical feasibility of the invention is not confined to a particular plant or animal variety;

(c) a microbiological or other technical process, or a product obtained by means of such a process other than a plant or animal variety.

Rule 28: Exceptions to Patentability

(corresponds to Art. 6(2) EU Directive)

Under Art. 53(a), European patents shall not be granted in respect of biotechnological inventions, which, in particular, concern the following:

(a) processes for cloning human beings;

(b) processes for modifying the germ line genetic identity of human beings;

(c) uses of human embryos for industrial or commercial purposes;

(d) processes for modifying the genetic identity of animals which are likely to cause them suffering without any substantial medical benefit to man or animal, and also animals resulting from such processes.

Recital (41) of the Directive 98/44/EC defines a process for cloning human beings "as any process ... designed to create a human being with the same nuclear genetic information as another living or deceased human being". Any manipulation of a human cell or gene aiming to produce a human being, whose genome is not composed solely of unmodified genes from *both* his/her father *and* mother, is not patentable. This exception of course does not apply to the preparation of human tissues or organs if this preparation is not aiming to produce a human being. It is, however, unlawful to patent "an invention aimed at isolating from its natural state an organ of the human body, for instance a kidney, in order to sell it" [19]. In this context it should be noted that the exceptions for patentability in Rule 28 are merely to be considered as examples of non-patentable subject matter. Thus, developments not explicitly mentioned in Rules 26 to 29 are not necessarily patentable.

These rules only give a guideline of what the European States consider as not being patentable. For instance hybrid organisms displaying features of animals and human beings are certainly not patentable [20] due to the general stipulations in Article 53a EPC although not explicitly mentioned in Rule 28. Not patentable are also processes for modifying the genetic identity of the germ line of human beings. The germ line begins, according to the understanding in patent law, with the fusion of the gametes that means the oocytes and the sperms. Thus, processes aiming to change the genetic identity of somatic cells or aiming to manipulate merely the gametes without intending to change the germ line appear to be patentable. This appears to apply also to uses of *human* embryos for non-industrial and non-commercial purposes such as therapeutic or diagnostic purposes, which are applied to the human embryo and are useful to it (Recital 42; according to the official German interpretation the term "useful to it" means to the very same embryo, i.e. not a foreign one—conf. Sect. 4.5—whereas the interpretation of some other countries, such as Denmark, France, and Sweden, include the usefulness to other human embryos). Rule 28 item (d) reflects the situation of the Oncomouse/HARVARD-case discussed below (Sect. 4.2) and tries to balance out the chances and risks of technologies involving mankind and genetic modifications in animals.

Rule 29: The Human Body and its Elements

(corresponds to Art. 5 EU directive)

1. The human body, at the various stages of its formation and development, and the simple discovery of one of its elements, including the sequence or partial sequence of a gene, cannot constitute patentable inventions.
2. An element isolated from the human body or otherwise produced by means of a technical process, including the sequence or partial sequence of a gene, may constitute a patentable invention, even if the structure of that element is identical to that of a natural element.
3. The industrial application of a sequence or a partial sequence of a gene must be disclosed in the patent application.

Rule 29 clarifies that the human body and its elements per se are not patentable. The human body in all of its different developmental stages is excluded from patent protection, which includes the very first part of human life that means the fertilized egg. The human body finalizes, at least according to the prevailing literature opinion, its existence with its death. Accordingly, a dead human body may be subject to patent protection, of course always under consideration of Article 53a EPC as explained above. Rule 29 further stipulates that a human body can be the subject of patent protection in the form of elements thereof if these elements constitute an invention in contrast to a mere discovery. Accordingly even elements from the human body, which are identical to the elements, found in or on the human body are patentable if the essential first criterion for patentability is fulfilled, namely that the element is not a discovery but an invention, that means a technical teaching.

It must be realised that the above stipulations in Rules 26 to 29 EPC are of clarifying and additive nature to the general criteria for patentability discussed above. In particular and in addition, each single invention must be new and inventive to be patentable.

3.4 Pitfalls on the Approach to Valid Patents

Law, rules and instructions for the examiners regulate the patenting process. Still, errors or blunders may sometimes creep into the examination procedure. Regarding the novelty requirement, for instance, inventions can be patented if the examiner is not aware of any information rendering the invention known in the art. In spite of all modern information retrieving systems, neither an applicant for a patent and nor a patent examiner can be fully aware of every piece of prior art relating to the particular invention. Another source of errors may arise from underrating the scope of the patent. As an example, a patent applicable to and *expressis verbis* intended for animals may silently be technically applicable also to humans just because this application was not explicitly excluded in the description and/or the claims.

Day for day patent authorities all over the world have granted, are granting and will grant patents which are totally or in part not legally valid since not all relevant facts could be assessed properly. However, the patent systems provide means to revoke or restrict legally invalid patents such as opposition and invalidity proceedings. These means serve to tailor the patents granted into legally valid patents that are in line with the correct and full scope of the patent stipulations. Thus, the pure existence of a legally invalid patent is of no threat to the public or a particular third party. Because of the fact that such a patent is invalid because of conflicting with any of the above-identified patentability criteria, it cannot be validly used against a third party.

On the other hand, annulations of a patent or of single claims may be an expensive matter for the plaintiffs, in particular if the suits pass through several instances.

4 Milestones in Patenting Biotechnological Inventions in Europe and Germany

4.1 The Development of Patenting Policy for Biological Material in Germany

Patenting a biotechnological invention was (and currently is) not always an easy thing to achieve. In fact, in 1877, when the first German Patent Act entered into force, the legislator had not even thought of biotechnological inventions as being the subject matter of patents. Patents have been originally designed to cover inanimate technical developments. From 1877 until the end of the 1960s, there was only a very slow movement from excluding patent protection for any biological invention to patenting at least the influence of humans on processes of living nature such as plants, animals or even human beings. Thus, Flemings observation that penicillin inhibited the growth of gram-positive bacteria in the 1920s, the use of various compounds including antibiotics against infections in the 1930s and 1940s and the biotechnological production of vitamins, amino acids and enzymes in the 1950s were not rewarded with patent protection. However, the scientific and technological development in the 1960s made more and more clear, that the biological living nature too is governed by and uses chemical and physical processes. Avery [21], Franklin, Watson, and Crick [22] founded the modern biotechnology by allocating heredity to chemical structures, namely DNA. Jacob, Monod [23] and Nirenberg [24] showed the way from DNA to the phenotype of living organisms. The German Federal Supreme Court (BGH) in its decision 1969, BGHZ 52, 74 on "Rote Taube" (red dove) [25], consequently ruled, still under the German Patent Act 5th of May 1936, that also biological processes can be controlled by man and are predictable. Thus, the planned use of biological forces and processes was considered to be susceptible to patent protection. In said decision, the BGH considered in principle a process of breeding animals being patentable, provided the process is repeatable. A first hindrance to patenting biological inventions was overcome. However, at that time the German Patent Act still contained a prohibition to grant patents for pharmaceuticals, food and confections as well as substances prepared by chemical synthesis.

With effect of the 1st of January 1968 the German Patent Act of 4th of September 1967 has repealed said prohibition. Subsequently, chemical substances prepared by chemical synthesis became the subject matter of patents. It was not long before the question arose whether chemical substances occurring in nature could also become patented. In the 1970s, it indeed became established case law, that naturally occurring substances are patentable, if they have been isolated and could be correlated to a technical application, even if said isolated substance has the same structure as the naturally occurring substance (BPatG "Antamanid", 16 W (pat) 64/75; BPatG E20, 81; BPatG "Menthonthiole" 16 W (pat) 81/77 [26–28]). Thus, the mere discovery or identification of a substance in nature is not an invention. The inventor who has provided to the public a hitherto not available substance, be it naturally occurring or chemically synthesised, for an interesting purpose, deserves a patent while substances freely occurring in nature stay free of any exclusive rights.

It again did not take a long time before the question came up, whether biological living subject matter could become patented. In 1975 the BGH granted substance protection on a microorganism, namely a yeast, provided the inventor exemplified a way to reproduce the invention, that means to reproduce the patented micro-organism (BGH, "Bäckerhefe" (bakers' yeast), BGHZ, 64, 101). The German Federal Patent Court granted in 1978 patent protection for another micro-organism, namely *Lactobacillus bavaricus* (BPatGE 21, 43).

In 1987 the BGH finally granted patent protection for a virus even if its reproduction in vivo was not guaranteed but under the provision that the virus had been deposited and is available to the public (BGH, "Tollwutvirus" (rabies virus), BGHZ, 100, 67). The breakthrough was achieved.

The possibility to deposit microorganisms as a supplementary part of the disclosure of a microbiological invention was confirmed by the German Federal Patent Court as early as 1967 and by the German Federal Supreme Court in 1975 [29]. The final step, in 1977, was the Budapest Treaty on the International Recognition of the Deposit of Micro-organisms. The implementing regulations to the European Patent Convention of October 1973 adopted the deposition rules from national patent legislation and clarified and unified the rules (Rules 31 and 33 to 34). In Germany the regulations for depositing micro-organisms is regulated by a decree of the GPTO [30].

The implementation of the Biotechnology Directive (Sect. 4.3) into German National Law will be treated in Sect. 4.5.

4.2 Biotechnological Patenting Problems as Faced and Handled by the European Patent Office

Parallel to the developments in German case law, on the 7th of October 1977 the EPC has entered into force for Germany and six further European countries. The EPC created, for the contracting European states, a common procedure to obtain a European Patent, which in each contracting state exerts the same effects, as a national patent would do. Thus, it provides for the contracting European states a common and harmonised granting procedure ensuring for the patent applicants in all the contracting states a predictable and harmonised patenting process. Furthermore, the EPC foresees a common protocol to interpret the terms of a patent in matters of patent infringement. The EPC grants patents by virtue of one of its organs, namely the EPO, which is responsible for the granting and subsequent opposition proceedings. Thus, the EPO is working independently of the still existing national patent offices, and from 1977 on developed its own case law on the patentability of inventions. This case law both has been influenced by the national case law and does it vice versa. Up to 2004, 18 countries have joined the EPC.

In the course of the rapidly developing field of molecular biology-based medicine, biotechnology and microbiology, the number of patent applications relating to biotechnological matters including subject matter relating to nucleic acid sequences, proteins, microorganism, transgenic animals and plants increased tremendously. One of these cases related to Leder's and Stewart's well-known "Oncomouse invention" (Patentee: President and Fellows of Harvard College) [31]. The patent in question relates to a transgenic non-human mammalian animal, whose cells contain an activated oncogene. Said oncogene caused the claimed mammalian animals, in particular mice, to develop cancer, which made them particularly useful for developing anti-cancer drugs. In its decision T 19/90, "Oncomouse/HARVARD" [32] the Board of Appeal held that the patentability exclusion in the EPC with respect to animal varieties does not rule out patents on animals in general. The Board remitted the case to the Examining Division, to examine and decide inter alia whether the invention in question would not be an invention the publication or exploitation of which would be contrary to "ordre public" or morality. The Examining Division in turn came to the conclusion that the Oncomouse invention was complying with the EPC indicating that the purpose of the patented teaching to facilitate cancer research and prevention was of such high importance for humanity as to outweigh any disadvantages, such as suffering of the animals concerned. After opposition by several organizations from Austria, Germany, Switzerland and the UK, on 6 July 2004 the Technical Board of Appeal (T 0315/03) confirmed the patent but narrowed its claims to *transgenic mice*, whereas the Examining Division as the first instance had the claims restricted to rodents instead of—originally—relating to "transgenic non-human animals". (For comparison, the related US patent [33] is still valid in its original scope.) The oncomouse patent was the first patent granted by the EPO for a transgenic animal.

In accordance with the stipulations in Article 53 EPC and with the decision T 19/90, it was the prevailing opinion, that not only animals in general, but also plants in general would be patentable as long as the classification of the plants or animals is higher than the variety level. It was quite a surprise when a Technical Board of Appeal decided in the case T 356/93, "Plant cell/PLANT GENETIC SYSTEMS" [34], that a broad claim relating to a transgenic plant [35] would embrace plant varieties and thus would not be patentable under Article 53 EPC. Because of its fatal consequence to the development of plant biotechnology in Europe, the decision provoked a dramatic response both in literature and practice. In this context it should be noted, that at that time about one thousand applications were pending in the field of patenting plants. In consequence to T 356/93, the Examining Divisions of the EPO rejected patent applications containing claims directed to plants. Considering that the rest of the world, namely the global players in the terms of economics maintained their practice of patenting plants, quite an unsatisfactory situation for the European applicants developed. It was overdue that finally by the decision G 1/98 ("Transgenic plant/NOVARTIS II" [36]), the Enlarged Board of Appeal of the EPO held that a claim wherein specific plant varieties are not individually claimed [37] is not excluded from patentability, even if it may embrace plant varieties. The decision clarified that plant varieties containing genes introduced into an ancestral plant by recombinant gene technologies irrespective of the way they were produced are excluded from patentability, as long as the plant variety itself is claimed. As a consequence of this decision, it became possible again to obtain claims directed to transgenic plants as long as they did not specifically relate to individualised varieties. This is in line with the EU directive (98/44/EG) discussed in Sect. 4.3 indicating that plant and animal varieties are not patentable but that inventions which concern plants or animals shall be patentable, if the technical feasibility of the invention is not confined to a particular plant or animal variety.

The development of case law by judiciary decisions is but one although important influence on patent legislation. In the course of parliamentary legislation, governments and legislative bodies have to consider the opinion of patent jurists, industrial experts, the scientific community, and non-government organisations. Legal advisers are called upon to provide solid and balanced reports and well-founded suggestions to the preparation of amendments to patent acts or statutes [14, 29, 38].

4.3
The European Unions Biotechnology Directive 98/44/EC

In the 1980s already, the European Parliament recognised that biotechnology and gene technology play an increasingly important role in various industrial applications. It was recognised, that an appropriate intellectual protection for biotechnological inventions would be of outstanding importance for the economic and scientific development of the European Community. It was furthermore recognised that an effective and harmonised protection would be desirable motivating for inventions in all areas of biotechnology. The European Parliament also adopted the view that only a unified legislative development would prohibit unfavourable effects on trade and industrial development. Such unfavourable effects were expected to possibly derive from divergent moral views relating to the application of biotechnology techniques. One of the most prominent aims when unifying the European law in this area was to exclude particular subject matter from patentability, if said subject matter does not comply with a unified ethical standard. After more than ten years intensive discussion, on the 6th of July 1998 the "EU directive 98/44/EC of the European Parliament and the Council on the Legal Protection of Biotechnological Inventions (Biotechnology Directive)" [18] was published and put into force. All the contracting states of the EU had to implement the directive into national law within two years of the date of the publication. EU directives are binding law for the contracting states, so that the states would have had to amend their patent laws accordingly, wherever necessary (Article 15 (1)). Up to 2004, Germany and seven other states (Austria, Belgium, France, Italy, Luxembourg, the Netherlands and Sweden) had not yet implemented the EU directive into national law. The Netherlands supported by Italy and Norway even filed an action of annulment against the EU-Directive at the European Court of Justice. Said action was rejected on the 9th of October 2001, confirming that the EU directive is in agreement with the EU ethical standards as well as with the relevant legal basis, and furthermore, that the EU directive is not in contradiction with any other international convention or obligation. The EU aiming at the same time to enforce the law and to control its effects, had decided to file a law suit against the above eight states at the European Court of Justice for not timely implementing the EU directive into national law while the European Council decided to closely survey the developments in biotechnology and patenting biotechnology. In particular, attention was to be given to the scope of protection for patents relating to nucleotide sequences isolated from the human body and to the patentability of human stem cells and cell-lines derived there from. As a matter of fact, the European Court already sentenced France as one of those eight countries in July 2004, and Germany was cautioned. In the meanwhile, the amendment to the German Patent Act of January 2005 [39] foresees an almost one-to-one implementation of the EU directive (cf. Sect. 4.5).

4.4 Consequences of the Biotechnology Directive for Patenting Biotechnological Inventions in the Contracting States to the EPC

Although the EU directive is not binding for the European Patent Organisation (the European Patent Organisation is not a member state of the EU), the implementing regulations of the EPC—which are the rules to implement the EPC—have been amended to adopt the EU directive. Consequently, effective as of the 1st of September 1999, the rules of the EPC were amended to incorporate new rules 26 to 29 EPC relating to the patentability of biotechnological inventions. The EU directive is therefore already reality in Europe.

It is meanwhile widely accepted, that the EU directive does not set out any new patent law for the contracting states. In fact, the EU directive is based upon case law of the various national contracting states and the EPO. Its importance must be seen in clearly defining subject matter and terms (Rules 26 and 27 EPC), in setting up a unified ethical standard exemplified by specific technologies complying and not complying with said standards (Rule 28 EPC), and, last but not least, by drawing a clear-cut line between a discovery and an invention (Rule 29 EPC). Drawing such a borderline became necessary due to the increasing number of patent applications relating to nucleic acids isolated and sequenced in the course of the various genome mapping and sequencing projects (e.g. HGP and Celera projects) around the world. It became established case law that patenting of human nucleic acids cannot be considered as intrinsically unethical (Opposition Division, "Relaxin") [40, 41]. In said decision it is held, that patents for nucleic acid sequences do not confer rights relating to individual human beings; thus, patenting of for instance DNA-sequences does not patent life per se, but solely isolated nucleic acid molecules. Completely in line with the older German case law on naturally occurring substances, it is also well-established case law, that DNA-sequences isolated from natural sources do not represent a discovery ("Relaxin", see above). In the decision "Novel V28 Seven Transmembrane Receptor" (Opposition Division) [42, 43] it is clearly stated, that a purified and isolated nucleic acid molecule comprising the naturally occurring nucleotide sequence does not exist in nature (in the isolated and purified form), and thus cannot be discovered. Consequently, a purified and isolated polynucleotide is not a discovery. However, to qualify as an invention, such a purified and isolated polynucleotide must fulfill the criteria of being a technical teaching, i.e. solving a problem with technical means, and being of industrial application. The polynucleotide therefore must be associated with a function that provides a meaningful technical teaching to mankind. Accordingly, the provision of a nucleotide sequence or a protein without a credible function does not constitute an invention. On top of these criteria, a nucleotide molecule or protein of course only can be patented if it fulfills the further requirement for patentability such as novelty and inventive step. Both criteria are not easily overcome. With the same speed as isolating, identifying, sequencing nucleotide molecules and allocating functions to them becomes routine experimentation, the room for inventive step disappears. Presently, it is much more difficult to argue for the presence of inventive step of a nucleotide molecule than in the 1980s or 1990s. This is not without reasoning: If an automated sequencer and a software-based homology search and analysis accomplish an invention there must be something wrong with the definition of an invention.

Among the many patents granted after the incorporation of the EU Directive in the implementing regulation of the EPC, several gained particular attention in the public and confirmed the necessity of applying a unified ethical standard to protect inventions. EP 0 578 653 B1 (so-called "Seabright-patent") [44] relates to the production of a transgenic fish comprising a chimeric gene of non-human origin. This patent relates to a transgenic fish and not a fish variety. Accordingly, the exclusion of patentability for animal varieties does not apply here. It is important to realise, that the chimeric gene used to produce the transgenic fish does not aim to produce a hybrid organism made from totipotent human and animal stem cells or even germ line cells. In such a case, the invention would relate to cloning of a chimeric organism of partially human origin, which would not be patentable.

EP 0 695 351 B1 (the so-called Edinburgh-patent) [45] relates in its granted version to the isolation, selection and propagation of animal stem cells. The description of this patent pointed out, that also human cells must be considered as being animal cells. Fourteen different opponents, among them various political and ethically oriented organizations, objected to this patent. In a preliminary decision of the Opposition Board of the EPO, which dealt with this case, the patent was maintained in amended form after the patentee had specified in the patents description that the animal cells are non-human animal cells. This patent appears now to be in line with the currently applicable European law [46]. It is noteworthy, for comparison, that the corresponding US patent 6 146 888 [47] remained in its original form, i.e. it comprises also human cells inclusive embryonic stem cells.

However, it will be interesting to follow up the forthcoming developments, both in the terms of legislation and case law. Are biotechnological inventions so different to other inventions as to necessitate their own law?

4.5 Implementation of the Biotechnology Directive into National Law in Germany

The main reason why several member states of the European Union failed to meet the deadline for implementing the Directive 98/44/EC into national law, was the feeling that the scope of protection provided by a product patent on genetic material was too broad. After long disputes both in the parliament and in public, the German legislature finally decided to amend the German Patent Act (Promulgation January 21, 2005, in force since February 28, 2005 [39]). In the following, the major changes are indicated *in italics*:

GPA §1a: Patentable Inventions

1. The industrial application of a sequence or a partial sequence of a gene must *specifically* be disclosed in the patent application *by specifying the function to be performed by the sequence or partial sequence.*
2. *If the subject of the invention is a sequence or partial sequence of a gene, the structure of which is identical to the structure of a natural sequence or partial sequence of a human gene, then its use according to the commercial applicability as described in section (3) has to be incorporated into the patent claim.*

Subparagraph (3) emphasises the importance of a function of a (partial) gene sequence for its patentability. The amendments in subparagraph (3) essentially reflect what already has been stipulated in Art. 5, item (3), together with recitals 23 to 25 of the Biotechnology Directive [18]. Thus, for the patentability of a sequence of a gene it shall not suffice to describe a general aim of an invention as for instance "DNA sequence useful for medical purposes". This may be sufficient to establish industrial application of a sequence, but does not suffice to establish a specific function of the sequence as required in § 1 (3). Thus, due to subparagraph (3), one has to distinguish more clearly between industrial applicability, which may be defined in broader terms, and a specific function of a sequence, which obviously represents a specific chemical or biological function.

Subparagraph (4) exceeds the wording of the biotechnological directive insofar as it now requires to incorporate the use according to the commercial applicability as described in subparagraph (3) into the patent claim. Since the scope of a patent is primarily determined by the elements of the claims, the sequences are protected only in so far as they are at least able to perform the indicated function. This requirement, however, solely applies for sequences of a gene, the structure of which is identical to the structure of a natural sequence of a human gene. Thus, this stipulation only applies to very specific cases, i.e. gene sequences that are identical to human gene sequences.

GPA §2: Exceptions to Patentability

(Corresponding to Rule 28 EPC and to Art. 6(2) EU Directive) contains an additional paragraph in order to define what the standards should be in applying the relevant EPC and EU rules referring to human beings (i.e. no processes for cloning nor for modifying the germ line genetic identity, and no uses of embryos for industrial or commercial purposes):

"*Upon the application of it. 1 to 3, the standards are set by the applicable provisions of the Embryo Protection Law: http://www.bundesrecht.juris.de/bundesrechteschg/index.html.*"

According to Recital 42 of the Directive the uses of human embryos are not excluded if inventions for therapeutic or diagnostic purposes are concerned, which are applied to the human embryo and are useful to it. The German government's official interpretation is that alien utilization ("Fremdnützigkeit") is excluded, but there remains an uncertainty about how this has to be interpreted: alien to human embryos or alien to the selfsame embryo to whom said therapeutic or diagnostic invention is to be applied?

GPA §2: More Exceptions to Patentability

(1) Patents shall not be granted in respect of plant and animal varieties or essentially biological processes for the production of plants and animals.
(2) Patents shall be granted in respect of inventions,
   1. the subject of which are plants or animals, if the realisation of the invention is technically not restricted to a specific plant or animal variety,
   2. the subject of which is a microbiological or another technical process or a product obtained by such a process *if no plant or animal variety is concerned.*

§ 1a it. 3 holds correspondingly.

Inventions of microbiological processes for the production of novel plants or animals can only be patented if their result is not a plant or animal variety. In contrast to Article 4 it. 3 Biotechnology Directive, GPA §2 it. (2) 2 excludes the patentability of plant and animal varieties even if they are obtained by microbiological processes.

GPA §9c, it. (3)

(7) § 9a it. 1 to 3 does not pertain to biological material, which in farming has been obtained accidentally or unavoidably by technical means. Hence, as a rule, no claim can be laid on a farmer if he has cultivated seeds or has bred plants, which are not subject to that patent protection.

§ 9a it. 1 to 3 correspond to Articles 8 and 9 Biotechnology Directive. The patent protection shall be restricted in accidental or technically not avoidable cases where accidental crossbreeding in farming occurs. Good agricultural practice is considered as the standard. The farmer shall be protected from an "imposed enrichment". In case of violation, the burden of proof is on the patent proprietor.

GPA §11 [Permissible activities], items 2 and 2a

The effects of a patent shall not extend to ...
2 acts done for experimental purposes which relate to the subject matter of the patented invention.
2a. *the exploitation of biological material for the purpose of reproduction, inventing and developing a novel plant variety.*
Item 2 (old) constitutes a research privilege. Item 2a refers to the German Delegations statement in the minutes of the Internal Market Council of 27th November 1997, footnote 2, requiring that "the breeding of plant and animal varieties shall not unduly be impaired by the effect of patents for biological material".

GPA §34a

If the subject of an invention is biological material of plant or animal origin or if such material is being used in that context, then *the application shall contain statements about the geographical place of origin of that material insofar as it is known. The examination of the applications and the validity of the rights on the basis of the granted patents is not affected by this.*

This amendment corresponds to Recital 27 Biotechnology Direction and intends to make things transparent without anticipating the results of the relevant Expert Committee of the World Intellectual Property Organisation. It takes into account the Convention of Biological Diversity, which cares for equal access to genetic resources and benefit sharing.

5
Some Remarks on Objections against Patenting in Biotechnology

Modern biotechnology has faced objections and controversy since the public awareness of gene technology, cell and molecular biology. In this section we shall not dwell upon the controversies about biotechnology as such, but restrict our report to some typical arguments against *patenting* in biotechnology.

In the opposing literature, different kinds of objections against "biotech patenting" have been advanced: ethical scruples as well as fears concerning monopolization of modern food and medicinal drugs, alleged impediment of research and development, or the dread of enlarging the prosperity chasm between rich and poor countries.

5.1 Ethical Aspects

As elucidated in Sect. 2, an essential item of patent legislation is to exclude, for a given time, anyone except the patent proprietor from exploiting the protected invention. Under this aspect, the patent legislation would not need to refer to general legislation and ethical aspects, which are to be obeyed anyway [50]. Nevertheless, European patent regulations as well as national patent acts contain ethical restrictions and rules to protect human dignity and animal rights. Consequently, many ethical arguments put forward by opponents to patenting in biotechnology are already embodied in patent legislation. Other ethical arguments are taken into account by the general legislation or other European and national acts protecting and ensuring dignity, privacy and integrity of all aspects of life. Furthermore, it is not existing life that is patented, but technical inventions in connection with living organisms and their products. If the invention concerns a gene modification, which is expressed and exerts its desired effect in a whole organism, the patent protection necessarily has to comprise the whole organism (e.g. the oncomouse) for the lifetime of the patent.

5.2 Scope of Patent Protection and Situation of Dependent Patents

Several objections demand that isolated genes or gene sequences—as carriers of information—should not be patentable as substances per se, even if the isolation involves an inventive step and if industrial application is given (Article 57 EPC). A gene sequence, it is argued, can contain information for several proteins (multi-functionality of genes) and therefore may be used for different applications not envisaged by the original inventor, who isolated the gene sequence and allocated for example solely a very specific therapeutic use to it. Thus, conventional substance protection, which is absolute and covers each and any use of the substance—even non-therapeutic uses—would not be justified. A further inventor finding a second application (say e.g. a diagnostic method) for the (same) patented gene sequence would in principle have the chance to obtain a patent on the invention but would be frustrated by the dependence of his patent on the former patent. This dependence of the second patent on the first one, which covers the gene sequence, opponents maintain, would inhibit medical drug development.

These arguments would only hold if the second (and further) invention(s) would overlap in *essential* parts of the DNA sequence in question, because Recital 25 Directive 98/44/EC rules:

Recital 25

"Whereas, for the purposes of interpreting rights conferred by a patent, when sequences overlap only in parts which are not essential to the invention, each sequence will be considered as an independent sequence in patent law terms."

In combination with **Article 5 (3)**, which corresponds to Rule 29 (1) Implementing Regulations to the EPC as already cited in Sect. 3.3—"*The industrial application of a sequence or a partial sequence of a gene must be disclosed in the patent application*"—the Government official statement to the Bill of the Federal Government for the Implementing of Directive 98/44/EC into German Law [39] defines: "By means of the description of the function, the patent examiner has to restrict the patent to that part of the gene applied for, which is essential for the function described, and has to exclude from patent protection those gene sections, which were applied for but are not needed for the function. Therewith the problem of sequence overlaps (cf. Recital 25 of the Directive) is largely avoided."

However, if there is an overlap, **Article 9** of the Biotechnology Directive applies: "The protection conferred by a patent on a product containing or consisting of genetic information shall extend to all material, save as provided in Article 5(1), in which the product is incorporated and in which the genetic information is contained and performs its function."

Then, in the case of overlap of essential parts of the sequence, a usual alternative can be (cross-) license negotiations, or, if necessary and justified, an application for a compulsory license (Recitals 13, 52, 53 and Article 12 Directive 98/44/EC; § 24 German Patent Act). The risk that such endeavours can fail would not justify abolishing the patent protection for substances. The risk for the second inventor is the chance and spirit for the first inventor. On the other hand, the dependence of such a product patent might as well wake the ambition to find other biotechnological paths for drug development. Thus, one possible way out would be to restrict the scope of the patent protection to the substance under the proviso that the substance is able to function in the way the (first) inventor envisaged. Another, even more restrictive way foresees to confine the scope of protection to the specifically invented use, thereby abolishing completely the "classic" substance protection as known for chemicals since decades. This is exactly what was realized in Germany (cf. Sect. 4.5) and in some other European countries —such as France, Switzerland, Spain, Portugal, and Italy.

However, when limiting the scope of protection for some kinds of inventions (for example biotechnological inventions), we have to realize that we are not alone in the world. International agreements and conventions on Trade and Intellectual Property [51] foresee a minimum standard for patent protection for any invention, regardless of the technical nature of this invention (Art. 27 TRIPS [52]). Thus, a national solution to the above problems should consider the international environment. On the other hand, the national decisions of the aforementioned countries exert a pressure on the European Community to change the Biotechnology Directive towards a more restricted product protection for isolated human genes and gene sequences. Some experts expect that the Biotechnology Directive will be adjusted to those national developments.

The future development will show whether the decreasing probability for inventing novel isolation methods for DNA sequences (cf. Sect. 4.4) would take the ethical and economic bite out of this problem.

Article 10 of Directive 98/44/EC holds consequences for plant breeders and farmers:

Article 10

"The protection referred to in Articles 8 and 9 shall not extend to biological material obtained from the propagation or multiplication of biological material placed on the market in the territory of a Member State by the holder of the patent or with his consent, where the multiplication or propagation necessarily results from the application for which the biological material was marketed, provided that the material obtained is not subsequently used for other propagation or multiplication."

Farmers may use protected seeds they bought for sowing and harvesting but—except for the use on their own farm (Article 11)—are not allowed to retain seeds from the harvest for selling or gifting those to third parties. The "farmers' exemption" or "farmers' privilege" is regulated by article 11 Directive 98/44/EC:

Article 11

1. By way of derogation from Articles 8 and 9, the sale or other form of commercialisation of plant propagating material to a farmer by the holder of the patent or with his consent for agricultural use implies authorisation for the farmer to use the product of his harvest for propagation or multiplication by him on his own farm, the extent and conditions of this derogation corresponding to those under Article 14 of Regulation (EC) No 2100/94.
2. By way of derogation from Articles 8 and 9, the sale or any other form of commercialisation of breeding stock or other animal reproductive material to a farmer by the holder of the patent or with his consent implies au-

thorisation for the farmer to use the protected livestock for an agricultural purpose. This includes making the animal or other animal reproductive material available for the purposes of pursuing his agricultural activity but not sale within the framework or for the purpose of a commercial reproduction activity.

3. The extent and the conditions of the derogation provided for in paragraph 2 shall be determined by national laws, regulations and practices.

The German Patent Act of January 28, 2005 adds an extra regulation in its §9c it.3 (cf. Sect. 4.5).

In contrast to the farmers' privilege of Article 11, the patent legislation does not grant a "breeders' privilege" as provided by Variety Protection Acts for "natural production" —such as crossing and selection [53]. The Variety Protection Legislation, however, is not a topic of this report.

Occasionally, it is alleged that patenting would hamper basic science. However, patent dependency does not necessarily affect basic science, since patent protection does, at least in Germany, not cover experimental research work done on a patented subject matter itself. The German Patent Act emphasizes this situation by formulating an experimental use exemption privilege (cf. Sect. 4.5, § 11 it. 2 and 2a; cf. Sect. 4.5).

As already mentioned in Sect. 2, patents force inventors to further scientific and technical progress by publishing their inventions instead of keeping them secret.

5.3
The Particular Needs of Developing Countries

Another set of arguments refers to the situation in developing countries and least-developed countries, which would not have the financial means to pay for patented medicines or for transgenic animals or food crops and their seeds. This should not be an argument against patenting of biotechnological inventions, but it will remain a matter of discussion for some time [51]. Such Third World problems are taken care of in Recital (11), Biotechnology Directive 98/44/EC [18]:

"Whereas the development of biotechnology is important to developing countries, both in the field of health and combating major epidemics and endemic diseases and in that of combating hunger in the world; whereas the patent system should likewise be used to encourage research in these fields; whereas international procedures for the dissemination of such technology in the Third World and to the benefit of the population groups concerned should be promoted ..."

By way of compulsory—possibly gratis—licenses or limitations of patent rights, patent systems all over the world are in principle able to take care of providing a balance between the rights of patentees and the economical framework of developing countries.

For instance, the German patent regulations allow in cases of well-founded requests to use patented biological material in the public interest under a compulsory license (§ 24 German Patent Act, similar regulations in other national patent acts), or, under certain circumstances, provide for compulsory licenses for plant breeders (Article 12, 1 and Recitals 11 and 13, Directive 98/44/EC). On an international basis the WTO agreed on special licensing regulations on the basis of the TRIPS Agreement [52], with alleviations for the least-developed countries (and in case of some other exceptions) by the Doha Declaration (2001) [54] and by the Decisions of the General Council for TRIPS of 27 June 2002 [55] and of 30 August 2003 (Cancún Ministerial) [56]. International negotiations about such exemptions for poor countries are going on in order to find a better compromise between the least-developed countries' need for payable medicines and the pharma and biotech companies' necessity to get a return for their R&D expenditures and, furthermore, to prevent the misuse that cheap medicines—instead of being distributed within the poor country—are imported or re-imported into the industrial countries at dumping prices. In addition to alleviations by international agreements and national legislative, financial support by national and international public agencies and non-government organisations is another way to provide patented medicines for poor countries.

In the field of green gene technology, a private initiative in line with those endeavours is an agreement between the inventors of GoldenRice [57] and six patent-holding companies to grant free licences to subsistence farmers (in poor countries) whose annual income on GoldenRice does not exceed US$ 10 000 [58]. GoldenRice is a genetically modified form of rice in order to fight vitamin A deficiency by an enhanced provitamin A content [59, 60] (and possibly other micronutrients etc in the future [61]). It has been tested in a field trial in Louisiana in 2004 but has not yet been admitted so far. It will be instructive to see how this voluntary agreement will be realized in the future.

Another concern refers to biological material originating from developing countries, e.g. agents for medical drugs as contained in tropic plants. To some extent this is accounted for by GPA § 34a (cf. Sect. 4.5).

6 Concluding Considerations

Patent legislation is one of the prerequisites for technical progress, and this is true for biotechnology as well. Patents impose an obligation to disclose inventions in return for the privilege to exclude others from exploiting the invention for a limited time. The alternative would be complete secrecy and abandonment of scientific exchange for a much longer time than the relatively short period between the acts of inventing and of patent application. That is the philosophy behind patent regulations, and that is the experience of many scientists and engineers in the biotechnological fields.

Both biotechnology and patent law are highly sophisticated disciplines. Understanding the combination of both requires considerable effort and obviously raises problems when trying to communicate chances and risks of biotech patenting to the public. It is part of the biotechnologists' responsibility toward the public to shed a very clear and objective light on this situation.

It is important, hence, that biotechnologists use their influence to inform the public and to de-emotionalise the discussion. This includes making oneself and others aware

a) of the patent law essentially being a prohibitive law: "a patent for invention does not authorise the holder to implement that invention, but merely entitles him to prohibit third parties from exploiting it for industrial and commercial purposes; whereas, consequently, substantive patent law cannot serve to replace or render superfluous national, European or international law which may impose restrictions or prohibitions or which concerns the monitoring of research and of the use or commercialisation of its results, notably from the point of view of the requirements of public health, safety, environmental protection, animal welfare, the preservation of genetic diversity and compliance with certain ethical standards" [62];
b) that this prohibitive effect can refer—for a limited time—also to *bio*technological inventions, i.e. technical inventions in biotechnological systems or products, and in some cases technically modified organisms;
c) that the patent regulations have developed—and are still developing—in consideration of the case law, and hence can, by amendments, be improved and adapted to the development in technology and general legislation, also under ethical considerations. As an example, the former unspecified patenting of isolated DNA sequences (i.e. without disclosing a technical application) was stopped by Rule 29 EPC and § 1a German Patent Act of 2005 [39];
d) of the difference between social obligations concerning Third World countries and the responsible mission of patent legislation. The latter one cannot adequately fulfill the former task, nor should it—apart from some well-defined exceptions.

It appears desirable, that bio(techno)logical, medical, biochemical and agrarian disciplines within the universities impart to their students information on, and a realistic and responsible attitude to patenting in biotechnology.

References

1. Pasteur L (1873) US Patent 141 072 (expired)
2. Chakrabarty AM, Chou G, Gunsalus IC (1973) Proc Natl Acad Sci USA 70:1137

3. Chakrabarty AM (1981) US Patent 4 259 444
4. Cohen S, Boyer H (1973) Proc Natl Acad Sci USA 70:1293
5. Cohen S, Boyer H (1973) Proc Natl Acad Sci USA 70:3240
6. Cohen S, Boyer H (1980) US Patent 4 237 224
7. Hammar F (2002) Adv Biochem Eng Biotechnol 75:1
8. Fiechter A (2000) Adv Biochem Eng Biotechnol 69:175
9. European Patent Convention 2000 as adopted by decision of the Administrative Council of 28 June 2001, (2003) OJ EPO, Special Edition No. 1
10. German Patent Act of 1981, last amended by the law of 29 August 2005
11. Pfaller W (2004) http://www.wolfgang-pfaller.de, last visited: Feb 2007
12. Pohlmann H (1960) GRUR 6:272
13. Straus J (1997) The present state of the patent system in the European Union—as compared with the Situation in the United States of America and Japan, Study in connection with the EC Commissions 1996 Green Paper on Innovation, EUR 17014 EN, Luxembourg; http://www.suepo.org/public/docs/2001/straus.pdf, last visited: Feb 2007
14. Straus J (2001) Grace Period and the European and International Patent Law. C.H. Beck, München
15. Trilateral Project B3b (2001) Mutual Understanding in Search and Examination, Report on Comparative Studies on Biotechnology Patent Practises, EPO, JPO, USPTO, San Francisco, Cal; http://www.trilateral.net/projects/other_project/business_method/ last visited: Feb 2007
16. Schulte R (2001) Patentgesetz mit Europäischem Patentübereinkommen, 6th edn. Heymanns, Köln, p 14
17. Implementing regulations to the European Patent Convention 2000 as adopted by decision of the Administrative Council of 7 December 2006
18. Directive 98/44/EC of the European Parliament and of the Council of 6 July 1998 on the legal protection of biotechnological inventions, (1998) OJ Europ Commun 30 July, L 213/13 EN
19. European Commission (2002) Report from the Commission to the European Parliament and the Council - Development and implications of patent law in the field of biotechnology and genetic engineering, p 17; http://europa.eu.int/eur-lex/en/com/rpt/2002/com2002_0545en01.pdf, last visited: Feb 2007
20. Recital (38), Biotechnology Directive 98/44/EC
21. Avery OT, MacLeod CM, McCarty M (1944) J Exp Med 79:137
22. Watson JD, Crick FHC (1953) Nature 171:737
23. Jacob F, Monod J (1961) J Mol Biol 3:318
24. Nirenberg M, Matthaei JH (1961) Proc Natl Acad Sci USA 47:1588
25. Bundespatentgericht (1969) XZB 15/67, GRUR 672
26. Sundt E, Ohloff G (1970) DE Patent 2 008 254
27. Sundt E, Ohloff G (1973) CH Patent 532 368
28. Bundespatentgericht (1978) 16W(pat) 81/77, GRUR 12:702
29. Straus J, Moufang R (1990) Deposit and Release of Biological Material for the Purposes of Patent Procedure. Nomos, Baden-Baden
30. Bundesgesetzblatt (2005) Biomaterial Hinterlegungsverordnung vom 28. Januar 2005, Tl I, Nr 2, p 151
31. Leder P, Stewart TA (1992) EP Patent 0 169 672
32. T 19/90 - 3.3.2 (1990) OJ EPO 476
33. Leder P, Stewart TA (1988) US Patent 4 736 866
34. T 356/93 - 3.3.4 (1995) OJ EPO 545

35. De Clerq A, Krebbers E, Vanderkerckhove JS, De Castro L, Gander E, Van Montagu M (1996) EP Patent 0 318 341 B1
36. Transgenic plant/NOVARTIS II (2000) OJ EPO 111
37. Hohn T, Peters C, Salmeron JM, Reed JN, Dawson JL (2000) PCT Patent Appl WO 00/60061; EP Patent 1 173 553
38. Beier FK, Crespi RS, Straus J (1985) Biotechnology and Patent Protection—An International Review. OECD, Paris
39. Bill of the Federal Government for the implementing of Directive 98/44/EC into German Law (2003); German Patent Act of January 28, 2005
40. Hudson PJ, Niall HD, Tregear GW (1984), EP Patent 0 112 149 B1; EP Patent 0 303 033 B1
41. Decision of the Examining and Opposition Division of 8 Dec 1994 (1995) OJ EPO 388
42. Godiska R, Gray PW, Schweickart VL (1998) EP Patent 0 630 405 B1
43. Decision of the Examining and Opposition Division of 20 June 2001 (2002) OJ EPO 293
44. Hew CL, Fletcher GL (2001) EP Patent 0 578 653 B1
45. Smith AG, Mountford PS (1999) EP Patent 0 695 351 B1
46. Zwicker J (2003) Mitt 11:502
47. Smith AG, Mountford PS (2000) US Patent 6 146 888
48. Krauss J (2005) Mitt 96:490
49. Einfinger A, Klein A (2006) Gen-Patente Pro und Contra, Humboldt Forum Recht, Beitrag 5, http://www.humboldt-forum-recht.de/5-2006/Drucktext.html, last visited: Feb 2007
50. Hartenbach A (2004) Die Bio- und Gentechnologie ist einer der Märkte der Zukunft, German Federal Ministy for Justice, Lecture on 11 March; http://www.bmj.bund.de/enid/Reden/Alfred_Hartenbach_zd.html?druck=1&pmc_id=1341, last visited: Feb 2007
51. Barton J, Straus J (2000) Nature 406:455
52. Agreement on Trade Related Aspects of Intellectual Property Rights (TRIPS) of 15 April 1994: Annex 1C of the Marrakesh Agreement establishing the World Trade Organization; http://www.wto.org/english/docs_e/legal_e/27-trips_01_e.htm, last visited: Feb 2007
53. Council Regulation (EC) No 2100/94 of 27 July 1994 on Community plant variety rights, Article 15 (c); Sortenschutzgesetz (German Plant Variety Act) of 1997/2001, § 10a (1)
54. Doha WTO Ministerial Declaration ("Doha Declaration") of 14 November 2001
55. Decisions of the General Council for TRIPS of 27 June 2002
56. Implementation of paragraph 6 of the Doha Declaration on the TRIPS Agreement and public health Decision of the General Council of 30 August 2003 (Cancún Ministerial)
57. Beyer P, Potrykus I (2000) DE Patent 19 909 637; PCT Patent WO 00/53768
58. Powell K (2006) Nature 24:294
59. Ye X, Al-Babili S, Klöti A, Zhang J, Lucca P, Beyer P, Potrykus I (2000) Science 287:301
60. Paine JA, Shipton CA, Chaggar S, Howells RM, Kennedy MJ, Vernon G, Wright SY, Hinchliffe E, Adams JL, Silverstone AL, Drake R (2005) Nat Biotechnol 23:482
61. Al-Babili S, Beyer P (2005) Trends Plant Sci 10:565
62. Recital (14), Biotechnology Directive 98/44/EC

Invited by: Professor Fiechter

Adv Biochem Engin/Biotechnol (2007) 107: 41–56
DOI 10.1007/10_2007_052

Published online: 28 February 2007

Bioscience, Bioinnovations, and Bioethics

Matti Leisola

Laboratory of Bioprocess Engineering, Department of Chemical Technology,
Helsinki University of Technology, P.O. Box 6100, 02015 TKK Helsinki, Finland
matti.leisola@tkk.fi

Dedicated to Elisabeth Fiechter, the late wife of Professor Fiechter

Abstract Biosciences and their applications, which we call biotechnology, have affected human society in many ways. Great hopes have been set on future biotechnology. The future depends on three key issues. First, we need good science. Recent developments in biosciences have surprised us in many ways. I shall explain in this article how. Secondly, we need structured innovation systems in order to commercialize our discoveries. Europe is slow in this respect compared to our Japanese and American competitors and may lose in the competition. I shall describe the Finnish innovation chain using the rewarded Otaniemi model as an example of how commercialization can be done in a systematic way. Thirdly, we need norms to guide what to do and where to go. Bioethics is probably the most neglected of the three key issues. With modern biotechnology we are able to do things that should worry every citizen, but the ethical discussion has been largely neglected or the discussion in our pluralistic society is leading nowhere. I shall finally discuss these problems from a historical perspective.

Keywords Biosystems · Biotechnology · Commercialization · Otaniemi model · Bioethics

1 Revolution in Biosciences

We tend to think that facts define our theory. Einstein realized that "it is the theory which decides what we can observe" [1]. The Austrian philosopher of science, Karl Popper, wrote in 1935 that scientists do not work according to the so-called scientific method [2]. The same was said even more sharply by Paul Feyerabend: "The attempt... to discover the secrets of nature and of man entails, therefore, the rejection of all universal standards and of all rigid traditions" [3]. Science is not a neutral field which is not influenced by philosophy, religion, or culture. Feyerabend claims that there is no systematic scientific method for revolutionary discoveries. Most new discoveries have been surprises and it has often taken a long time before they have been accepted by the establishment. It is fair to say that biology keeps on surprising us with its complexity.

1.1 Biological Surprises

Everything in biology is complex. This complexity has often been underestimated. Life was once considered a simple phenomenon which arose spontaneously from nonliving matter. Ernst Haeckel is known for his comparative studies on mammalian embryos and for his famous concept "ontogeny recapitulates phylogeny". Less well known is that he faked his embryo drawings [4]. Haeckel described the first living cell *Monera* which was supposed to be easily formed from nonliving matter, but he actually faked the *Monera* life cycle [5]. His *Monera* pictures were recycled in the literature for 50 years as the basis of the famous all-encompassing tree of the animal kingdom, although Pasteur's sterilization experiments had actually shown almost 10 years before that he was wrong at the start.

But now we know better! With the discovery of the genetic code and gene structure everything seemed clear. The central dogma was formulated: DNA $\rightarrow$ RNA $\rightarrow$ protein. This simplistic "gene-from-the-box" intuition is still in the minds of many of us. But biology keeps on surprising us. Firstly, the number of human genes is much lower than expected. We do not have many more genes than a mouse or *Drosophila*. Why are we so different? The number of genes seems not to be decisive. Secondly, eukaryotic genes seem to be fuzzy systems not so easily definable (Figs. 1– 4). One gene may produce several gene products via alternative splicing. Thirdly, introns that were considered useless contain microRNA which control gene expression. Now we know that one gene region can produce different transcripts which then can produce different proteins. Fourthly, even DNA code has proven not to be universal but variations of the standard code occur [6]. The surprises, however, do not end here. A few years ago much of the human DNA was considered to be

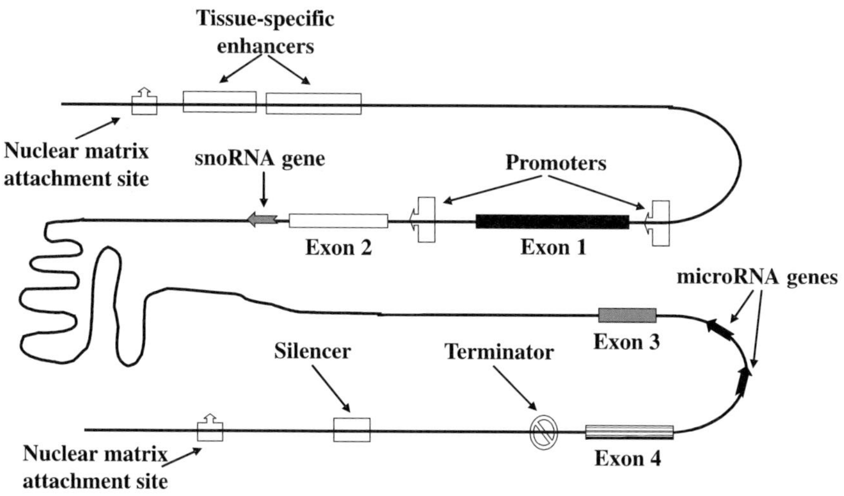

Fig. 1 Genes are concatenations of regulatory and protein-coding modules

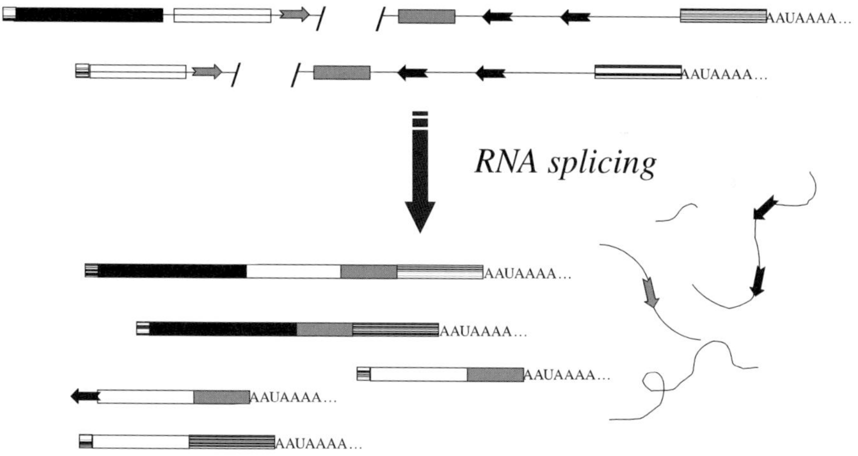

Fig. 2 Splicing of different primary RNA transcripts generates many gene products

composed of repetitive elements called junk DNA. This concept has now been largely overthrown when it became evident that these regions contain regulative functions [7].

One of the achievements of European biotechnology programs has been the sequencing of the yeast genome. The complexity of yeast has surprised us. For example, the large number of transport proteins for glucose on the cell surface was a surprise [8]. Then the cell content has proven not to be a chaotic flux of molecules but everything is highly structured. Large com-

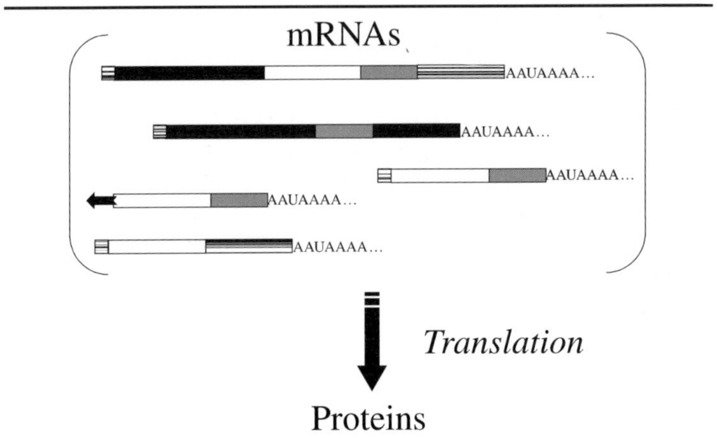

Fig. 3 Different RNAs are translated to many proteins

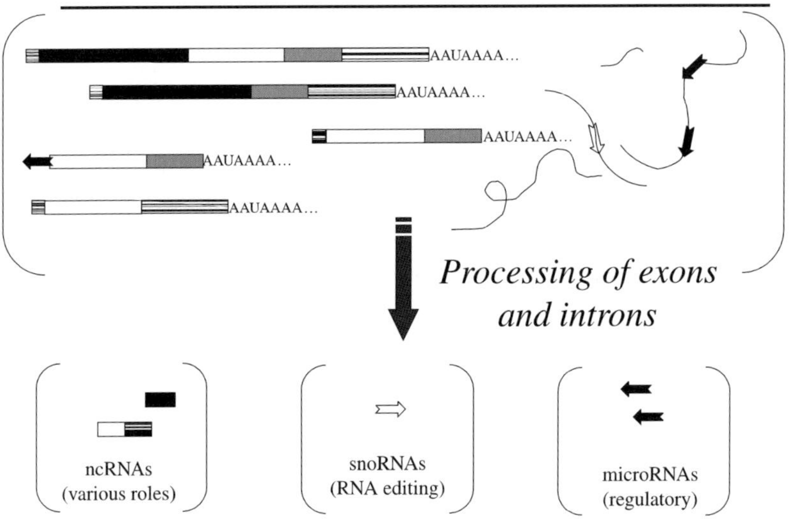

Fig. 4 Intron and exon RNA encode small regulatory RNAs

plexes of enzymes take care of transport of molecules from one enzyme to another effectively [9]. Even bacteria are not as simple as we thought: they have a complexity that is equivalent to that of eukaryotic cells [10].

1.2 From Reductionism to Systems

All these discoveries are revolutionizing our understanding of biology. The old reductionist view is being changed to a cybernetic systems biology view [11]. Table 1 summarizes the differences between the old atomistic model and the new interactive epi-genetic genome model (modified from [25]).

Biology is no longer studied only via individual molecules but as a dynamic system and we talk about proteomics, metabolomics, and molecular machines. Proteins interact with each other in a complex way [12]. Design, machines, and information are the metaphors for cell activities and the old Darwinian concept of randomness seems to be in trouble. A good example is the bacterial flagellum which has turned out to be the dream of nanotechnology. This rotary motor has been called the most effective engine in the universe [13]. We have no idea how it could have been formed by neo-Darwinian mechanisms. A new generation of scientists claims that nature is better understood as a product of design [14]. This view has created tremendous controversy recently [15].

The new view of genome complexity has affected very much the prospects of making rapid commercial innovations based on, for example, the known human genome sequence. The deterministic concept of one eukaryotic gene and one gene product is gone. So is also the concept of one gene causing one disease which can easily be treated with modern gene therapies or pharmaceuticals [16]. Scientists (including me) have long claimed that modern genetic methods are surgical tools to make only well-characterized and known changes compared to clumsy classical approaches. This view might be wrong. The system seems to determine what it does with the new genetic information, which means that the results of genetic manipulation may after all be surprises. Some have even suggested that the "phenotype overrides the genotype" [16].

I am sure that biology has not stopped to surprise us. In the early days of genetic engineering the hopes were high that we could engineer biologi-

Table 1 Conceptual change in biology from reductionism to systems biology [25]

Conceptual category	20th century atomistic model	21st century genome model
Scientific framework	Reductionism	Complex systems
Biological operations	Mechanical	Cybernetic
Central focus of hereditary theory	Genes as units of inheritance and function	Genomes as interactive information systems
Role of DNA	Passive vehicle of genetic information; active program during development	DNA as data-storage medium
Metaphor of genome organization	Beads on a string	Computer operating system

cal systems for our benefit fairly easily. The first successes, like production of human insulin in bacteria or yeast, supported this optimism. However, the road to commercial success has proven to be much more difficult than originally forecast. To understand the function and regulation of a cell on a more fundamental level we need a systems biology approach where biologists, mathematicians, and engineers work together.

2 Finnish Model for Commercialization

About 10 years ago I was a Scandinavian (and industrial) representative of a team which evaluated the EU 1990–1995 biotechnology program. A large amount of money was spent during that time. I was puzzled at how few commercial innovations were made. Only a couple patents were created. The major achievement was the sequencing of the *Saccharomyces cerevisiae* genome. The impression I had was that different research groups formed projects in order to be able to continue with what they were already doing. The extreme case was a plant biotechnology project with more than 100 participating laboratories. My understanding of a project is somewhat different. A project has a defined goal, timetable, budget, milestones etc. Basic science has a role, but when society invests billions in research it must have mechanisms to get some benefit from the investment. Otherwise the money would be better used elsewhere.

Finland is known for its technology and research-oriented culture. Here we can again use the word surprise. Nobody expected that a company manufacturing rubber boots and tires would become the world leader in mobile phones in a very short time. Now the name Nokia is known everywhere. In biotechnology Finns have been involved, for example, in creating one of the largest industrial enzyme companies called Genencor International. Other Finnish biotechnology innovations include the use of enzymes in animal feed, the pulp and paper industry, and fructose manufacturing. Finns have also been in the forefront of new brewing and dairy technologies.

Helsinki University of Technology (TKK) and VTT Biotechnology have been involved together with industry in many of these innovations. TKK is the largest and oldest of the Finnish technical universities. It is located in the Otaniemi peninsula about 10 km from Helsinki city center. In the same area is also the State Research Center (VTT) whose biotechnology unit is well known for its industrial biotechnology research. Central laboratories of the forest and brewing industries and the Geological Survey of Finland are situated in Otaniemi. Innopoli I and II offer facilities for new businesses and the Life Science Center offers space for more established companies. In biotechnology education, TKK works together with the University of Helsinki.

2.1 Support Tools

Finland has a national financing system for science-based innovations. The chain starts with the Finnish Academy (basic science), continues with the Technology Development Center TEKES (applied science), and finishes with capital investors like SITRA and BIOFUND. As in many other countries, biotechnology in Finland has been the focus area of research programs. However, the financing alone does not create new science-based businesses. The Otaniemi model of commercialization has been granted the Award of Excellence for Innovative Regions by the European Commission. The Otaniemi model (Fig. 5) has been developed to help innovators to commercialize their discoveries and ideas. Otaniemi International Innovation Center, which belongs to the Technical University, has had a central role in the model development.

The InnoTULI service is an important part of the Otaniemi Experienced Networking Model, especially in the evaluation of new ideas. It is the main tool in the cooperation between universities, research institutes, and science parks. InnoTULI grants small financing for the next-step development of business ideas. Details of this support tool are given in Fig. 6.

InnoTULI is part of the national TULI program. It is funded by the Technology Development Center of Finland. TULI funding is at maximum 10 000 € per project and it can be used to buy development services from consul-

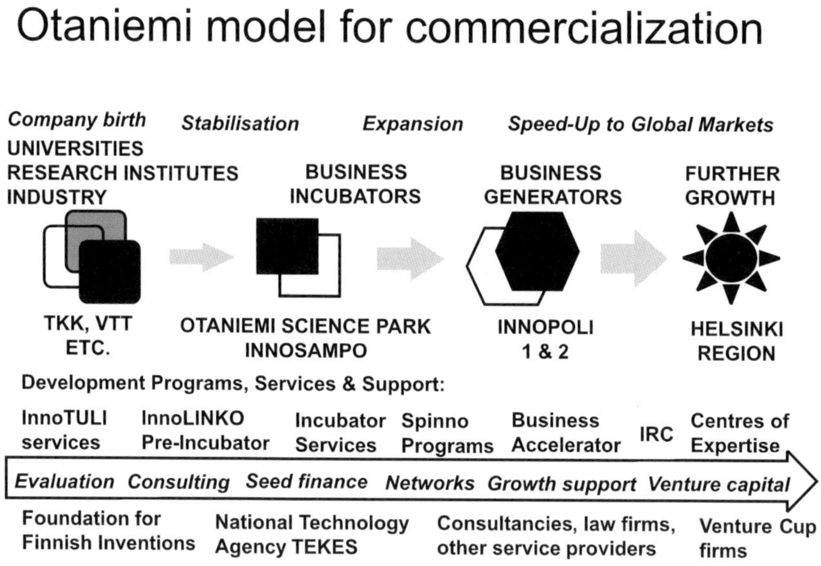

Fig. 5 Otaniemi Best Practice Experienced Networking Model

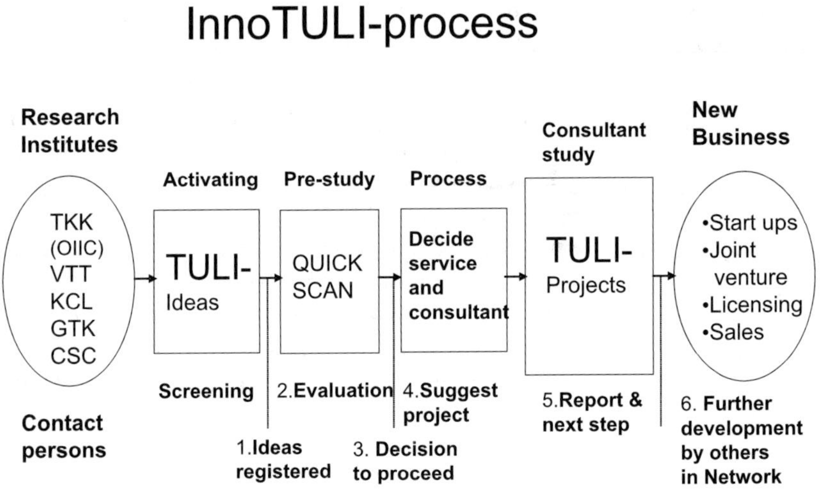

Fig. 6 Business exploitation of research results—Otaniemi Best Practice

tants or other service providers. Development services can, for example, be preliminary market or competitiveness analyses, preliminary business plans, partner search, novelty analyses, IPR issues, and other juridical or contractual issues. The Finnish network of science parks manages the TULI program.

InnoLINKO is a new early-stage business incubator of Otaniemi Science Park. InnoLINKO is jointly operated by Otaniemi Science Park Company and Helsinki University of Technology. It helps students and researchers to transform business ideas into new startup companies. It offers a concrete starting point for ideas emerging from research projects at TKK, as well as business plan and idea competitions like Venture Cup. The cooperation with the university enables the students to earn academic study credits while developing their own business.

InnoLINKO promotes a supportive atmosphere and an open communication between the teams. Tools for facilitating the sharing of knowledge and the exchange of experiences include InnoLINKO intranet and weekly roundtable discussions led by some of the teams or by a guest speaker. The companies can stay in InnoLINKO approximately 6 months in order to secure some form of pre-seed financing or to reach a stage in which subsequent growth relies on positive cash flow.

2.2 Active Search for Business Ideas

How to identify a good innovation? There is no patented answer to this question. There is no magic in business creation, no simple procedure to business success, and no place for unrealistic dreams. Realism, discussion, goodwill, and plenty of good luck are needed. Professional experts with a "sensitive nose" for business are needed for guiding and mentoring. A stimulating atmosphere and supporting instruments provided by the public sector are valuable. Expert services within the university to support academic spin-offs are available. More than anything else hard work is necessary! And then more hard work! And finally continuous hard work!

The Venture Cup competition promotes academic entrepreneurship. It was developed by McKinsey & Co in Germany in 1996 to mimic Silicon Valley networks. It includes a business plan competition with three stages: feedback, educational events, and coaching. The Finnish Venture Cup competition started in 2000 with 100 000 € prize money. Technopolis Innopoli acts as a partner by providing premises and a jury member.

2.3 Biotechnology Needs Long Development Times

My personal experience is with the enzyme industry. The Finnish Sugar Company (later Cultor Ltd., now Danisco), where I was the director of corporate research during 1990–1997, bought an old penicillin factory in 1979 and started production of starch enzymes in 1980. Ten years later the business was still not profitable and the company decided to join forces with Eastman-Kodak (later Eastman Chemical Company). The joint venture created Genencor International Incorporated in 1990. About 15 years after Finnsugar en-

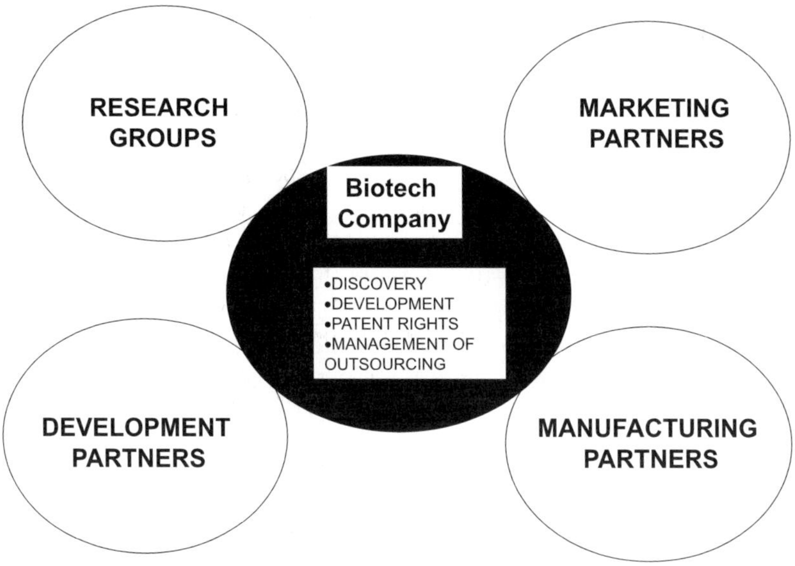

Fig. 7 Possible partnerships for a modern biotech company

tered the enzyme business the newly created joint-venture company was finally profitable. This involved the purchase of technology, production plants, and competitors as well as forming a joint venture.

The biggest expectations in biotechnology are, however, not in industrial but in pharmaceutical biotechnology. There the development times are even longer and the risks greater. Startup companies rarely have the possibilities to do everything themselves. Partnerships are usually necessary to speed up product development and for rapid market penetration. Typical partnerships of a modern pharmaceutical biotech company are shown in Fig. 7.

2.4
Public Perception

Public perception of biotechnology, especially in Europe, is somewhat negative. The experts have been trying to convince people that biotechnology bears no risks, is environmentally friendly, and will solve the future problems of mankind. This is of course an oversimplified picture. Many areas of biotechnology have been with us for centuries and everybody understands that without cheese, wines, vaccines, and antibiotics our world would be different. It is the manipulation of the hereditary material in plants and animals that causes the public to be concerned. Is this suspicion justified?

3
Ethical Considerations

Objectivity is the ideal of science. Truthfulness and honesty have therefore always been part of this idealism. Modern philosophy of science since the days of Popper, Kuhn, and Feyerabend does not, however, consider science as a neutral search for truth. Science has many faces. Modern biotechnology has raised questions that mankind has never faced before. We have learned to analyze, modify, and transfer genes. Many genomes have been sequenced and "pet cat cloning" raises excitement. Should we transfer genes between species? Is cloning of man acceptable? When does human life begin? Are only healthy people acceptable? Should we abort a human fetus if it is not healthy? Can we experiment with aborted fetuses? These questions show that the biosciences move in fields where the available answers are necessarily tied to the worldview of each individual.

The medicalization and geneticization are much more integrated in the "information society" than the 1930's eugenics legislation and compulsory sterilizations. Behavior-controlling drugs divert attention from counseling. Could the tools to sequence individual genomes be even more persuasive to the public to cure the recessive social traits than in the previous round of race hygiene? Human mental and moral traits are not "inherited in the same manner as hair color in guinea pigs". Somatic gene therapies and germline gene replacements are the key words for the future. Should the international insurance companies be permitted to deny coverage for families with preexisting genetic conditions? Can the insurance be made conditional on selective abortion? Genetic screening tests, amniocentesis during pregnancy, and dealing with deviant behavior are part of the future biotechnocracy.

3.1 Biotechnology Has Many Faces

The word "biotechnology" has at present different meanings in the minds of people. Many relate biotechnology to genetic engineering and pharmaceutical applications. Biotechnology is, of course, much more. The different areas of biotechnology have sometimes been defined by using different colors. *Red* biotechnology is related to human health and covers not only production of pharmaceuticals, vaccines, and biomaterials but also diagnostics and stem-cell research. *Green* biotechnology is related to food. It covers much of the traditional biotechnology and also new applications like the use of enzymes in baking and animal feed as well as genetically modified plants. *Blue* biotechnology has to do with the environment and includes purification of water, landfills, and exhaust gases. *White* biotechnology means production of chemicals like ethanol, acids, enzymes, vitamins, colors, intensive sweeteners, and polymers.

To discuss the ethical problems related to biotechnology I would like to group the research and development fields of biotechnology in a different way into four categories:

1. Classical breeding where microbes, plants, and animals have for a long time been developed by cross-breeding, selection, and traditional mutation techniques.
2. Modern production biotechnology where microbial, plant, or animal cell suspensions and enzymes are modified by molecular biology methods and used for production of various molecules.
3. Modification of plants and animals.
4. Genetic manipulation of humans.

3.2 Does Biotechnology Need Ethical Discussion?

Ethics tries to determine and justify norms (*ethos*) which can be used to direct and regulate biotechnical research, development, and applications. The future possibility of manufacturing cellulose, rubber, or spider's web protein by microbes does not seem to pose apparent ethical problems. But what should we think about products that can be used to eliminate human fetuses/embryos or as biological weapons?

The importance of ethical evaluation of any technology is proportional to the importance of the target of our activities. Norms are necessary when we start modifying nature by genetic methods. When the object is a human being, norms are mandatory. To clarify the point we can take an example. When a man by careless driving kills another human being he violates a more important norm than the one who accidentally kills a cat (although philosopher Peter Singer sees basically no difference between humans and cats). Because man is more important than a cat or cabbage, the norms that relate to treating humans are enormously important. The reputation of our bioscientific discipline is at stake. The norms are even more important when the manipulation target is the basis of our being—the human genome. Amputating a finger and manipulation of our genome are totally different things. Biotechnology dealing with manipulation of humans needs special norms because its destructive potential is huge.

A study of the history of science is vital for bioscientists to understand the need to exercise public self-criticism, when biology is being transformed from a descriptive and passive discipline to an active one. I have been studying with my colleagues the history of Finnish genetics [17]. The father of the Finnish eugenics legislation was Harry Federley (1869–1951). He corresponded with Ernst Haeckel and spent his critical years during the breakthrough from the recapitulationary paradigm to the Mendelism in Jena. Federley was a proponent of eugenics as were most scientists at that time. He lectured in Uppsala, Sweden, in the first race hygienic institute established in 1921, soon after the Finnish civil war with its rampant executions. He forced compulsory sterilization laws on Finnish society.

Federley's thinking goes back to the vulgar Darwinism popular in Germany and to its popularizer, Ernst Haeckel, whose influence on the racial views of Nazis and on the materialism behind Leninism was evident. A recent study of the activities of German scientists during the Nazi regime shows that "far from being subjected to force, many scientists voluntarily oriented their work to fit the regime's policies as a way of getting money and of exploiting the new resources. ... Most researchers, it turns out, seem to have regarded the regime not as a threat, but as an opportunity for their research ambitions" [18]. The situation at present is exactly the same but in a more subtle way. "Those who do not learn from history are doomed to repeat it" is not a just an empty phrase. Thus, the history of bioscience teaches us that we need ethical norms to guide our research. One can, however, ask why norms are needed in research and not only in applications.

3.3 Questionable Premisses

The ethical norms that direct our behavior are very much determined by our presuppositions (premisses). If our presuppositions are not valid then the norms based on them are without value and lack authoritative power. When we ponder the ethics of biotechnology the premisses are the most important thing. Therefore it is important to learn to recognize them in the ethical discussion. I shall give some examples.

The absolute value of science and the progress produced by science. There is an assumption that science always and automatically improves human society. The demand that science should have absolute freedom is, however, dangerous. Norms should guide our research targets and methods. (Joseph Mengele and his experimentation with humans is the most evident example of science without norms.) The negative consequences to future generations and accountability of our discipline should also be considered.

Neutrality of science and technology. It is erroneous to think that science and technology are neutral in themselves and ethics is needed only when the discoveries are applied. There is no science or technology that does not have a connection to the real world. Martin Heidegger understood that technology is a "construction" with its own internal dynamics which man cannot control. When a nuclear weapon has been developed the political forces sooner or later demand its use for "a good purpose". The story of Robert Oppenheimer from the Manhattan project that reached its "deadlines" quickly is a well-known tragedy. Every technology bears in itself built-in consequences. Corrupt individuals in a political system act usually as ice-breakers for questionable new technologies. This is accompanied by the interest of those who have been financing the research or the development of a technology (nobody finances for nothing). Finally, as Friedrich Nietzsche has pointed out, people have a natural tendency for power, and new technologies may give opportunities to seize power.

Ethical norms touch only politicians. It is wrong to assume that only politicians, who make decisions about the use of research results, need ethical norms. Institutions (state, universities, and companies) are not responsible for the application of research results but individuals (scientists, students, technicians, workers, and customers). They need norms to be able to detect which research results they should approve and which they should discard.

Utilitarianism sounds like a good premiss as a basis for ethical norms but it is problematic. How do we know what is good for mankind and nature in the long run? Oil catastrophes in Siberia, excess use of antibiotics, or diagnostics leading to selective abortion of baby girls are examples of this. It is even more difficult to give general answers to the question "What is good for a man?" when we do not even know who man is and why he ponders ethical questions, his own nature, and death. When we think of the complexity of natural systems it is questionable to think that we can master the consequences of human genetic manipulation. Utilitarianism actually means, in principle, that norms are relative.

3.4 Basic Premisses

Our premisses are determined by our worldview (basic premisses). When we consider genetic experimentation with humans the most important basic premiss is the right understanding of what a human being is. The materialistic Darwinian view of man is problematic. According to Darwinist philosophers Edward Wilson and Michael Ruse, ethics is "illusion caused by our genes" [19]. Richard Rorty says that there is no way of knowing whether social democrats are better than Nazis, modern medicine better than voodoo, Galileo better than inquisition [20]. The problem with the materialistic Darwinian view is that real discussion about ethics is meaningless because there is no proper place for right or wrong. Recent books like *Natural History of Rape* are examples of the present ethical dilemma [21].

The opposite of Darwinian ethics in the Western world has been Judeo-Christian ethics, which defines man as the image of God, having value since conception. According to Richard Weikart "many argued that by providing a naturalistic account of the origin of ethics and morality, Darwinism delivered a death-blow to the prevailing Judeo-Christian ethics, as well as Kantian ethics and any other fixed moral code" [22]. If our moral understanding is an expression of our evolving social instincts that change over time, the morality must be relative to the conditions we live in. Here we are at the most basic level of ethical discussion.

The decisive question dealing with human genetic manipulation is "Who is man?". If a human fetus is fully human, the norms touching him/her are radically different than if it is only a cell mass. The present situation is problematic in many ways. Professor Schlink gives the following example from Germany (*Der Spiegel*, 2003): an artificially conceived human egg is protected by law but when it is transferred to a womb it is no longer protected, and can be killed. Other European countries have similar legal problems. If a mother lets a newborn baby die she is guilty of murder, but if she decides to poison the baby a few months before the birth she has committed no crime. The idea of the sanctity of human life has disappeared and we can expect that more and more unborn, handicapped, and old people will be actively killed. In our pluralistic Western culture ethics is based on a majority opinion and the views that control are those of the ruling establishment supported by the mass media. I end this discussion with an example that reveals the present chaos in ethical and regulatory affairs.

My group developed a stable mutant of a microbial xylanase enzyme in 1999. The invention was patented [23] and the patent sold to an enzyme company. The stabilization was mainly a result of one disulfide bridge in the enzyme structure. We were able to show that all the properties of the mutant enzyme remained the same as in the wild type. The native enzyme is applied widely in animal feed in small quantities to help the animal digest its food better. Due to huge regulatory hurdles involving, e.g., toxicity tests, the product is, 7 years after its discovery, still not on the European market. Europe has such a regulatory jungle that we are automatically late with our research results compared to our competitors. The regulatory issues are, of course, often justified but not in this case. At the same time animals are mistreated during transportation, human fetuses are aborted without hesitation, experiments are being carried out with 14-day-old human fetuses, human cloning is being discussed, and experiments are even carried out to fertilize a human egg with animal sperm.

My personal view is that our ethical norms are floating in the air, since the Judeo-Christian rational roots as a basis for ethics have been largely replaced by relativistic situation ethics. There are no clear guidelines and the scientists can basically do what they want, and financial sources are considered "as an opportunity for their research ambitions" [17]. Naturalism has thrown supernaturalism out of any serious discussion and out of the science boat. This has even been considered as the most important discovery during the last thousand years [24]. Consequently the boat is without oars and rudder.

4
Conclusions

Individual nations and the European community are heavily investing in science and research. New discoveries, especially in biosciences, are continually made and they are changing rapidly our understanding of the complexity of biology. But how is the Otaniemi Science Park model to make business out of science related to ethical questions in biotechnology? The most effective systems can be the most destructive ones without norms to control them. Christian consensus in Europe and in the United States has given direction to society and science through the centuries. Modern science was born on Christian premisses in Europe. A reasonable God had created a reasonable world which a man, created in his image, can safely study and understand. The fathers of modern science, such as Kepler, Faraday, and Maxwell, sought to think "God's thoughts after Him."

Now when these premisses are mainly past history and there is nothing solid to replace them the ethical discussion has become "a lot of talk with little content". This is not so problematic when we want to produce chemicals with microorganisms. The problems come when we start manipulating human beings. Due to the materialistic worldview, which has replaced the Judeo-Christian understanding, humans are being dehumanized. We tend to think that the eugenic sterilization commonly practiced in the Western world up till the 1960s was a misuse of science. However, modern genetic technologies allow us to practice eugenics in a much more subtle way.

Acknowledgements I want to thank Richard von Sternberg for Figs. 1– 4, Veijo Ilmavirta for Figs. 5 and 6, Roland Tolksdorf and Pauli Ojala for ideas for the ethics discussion, and Ronald Nelson for language correction.

References

1. Popper K (1959) The logic of scientific discovery. Basic Books, New York
2. Heisenberg W (1971) Physics and beyond. Harper and Row, New York, p 63
3. Feyerabend P (1975) Against the method. Humanities Press, New Jersey
4. Gould SJ (2002) Nat Hist 3:42
5. Haeckel E (1868) Jenaische Zeitschrift für Medicin und Naturwissenschaft, Vierter Band. von Wilhelm Engelmann, Leipzig, p 64
6. Freeland SJ, Hurst LD (2004) Sci Am 22 March 2004
7. Shapiro JA, von Sternberg R (2005) Biological reviews. Cambridge University Press, Cambridge, p 1
8. Rolland F, Winderickx J, Thevelein JM (2002) FEMS Yeast Res 2(2):183
9. Ho Y, Gruhler A, Heilbut A, Bader G, Moore L, Adams S, Millar A, Taylor P, Bennett K, Boutilier K et al. (2002) Nature 415:180
10. Gitai Z (2005) Cell 120:577
11. Shapiro JA (2002) Ann NY Acad Sci 981:111
12. Giot et al. (2003) Science 302:1727
13. Berg H (2003) Ann Rev Biochem 72:19
14. Dembski W, Ruse M (eds) (2004) Debating design: from Darwin to DNA. Cambridge University Press, Cambridge
15. Brumfiell G (2005) Nature 434:1062
16. Silverman PH (2004) Scientist 24:32
17. Ojala PJ, Vähäkangas JM, Leisola M (2004) Jahrb Eur Wissenschaftskult 1:1
18. Editorials (2005) Nature 434:681
19. Ruse M, Wilson EO (1985) New Sci 108:50
20. Rorty R (1995) New Repub 31 July 1995:32
21. Thornhill R, Palmer CA (2000) Natural history of rape. MIT Press, Cambridge, MA
22. Weikart R (2004) From Darwin to Hitler: evolutionary ethics, eugenics, and racism in Germany. Palgrave MacMillan, New York
23. Fenel F, Leisola M, Turunen O (2002) Finnish patent FI108728
24. Mellman I, Warren G (2000) Cell 100:99
25. Shapiro JA (2002) Ann NY Acad Sci 981:111

Invited by: Professor Fiechter

Adv Biochem Engin/Biotechnol (2007) 107: 57–68
DOI 10.1007/10_2007_051

Published online: 28 April 2007

Genetically Modified Organisms in the United States: Implementation, Concerns, and Public Perception

Max P. Oeschger (✉) · Catherine E. Silva

Department of Microbiology, Immunology and Parasitology, Louisiana State University Health Sciences Center, New Orleans, LA 70112, USA
moesch@lsuhsc.edu

Abstract We examine the state of biotechnology with respect to genetically modified (GM) organisms in agriculture. Our focus is on the USA, where there has been significant progress and implementation but where, to date, the matter has drawn little attention. GM organisms are the result of lateral gene transfers, the transfer of genes from one species to another, or sometimes, from one kingdom to another. The introduction of foreign genes makes some people very uncomfortable, and a small group of activists have grave concerns about the technology. Attempts by activists to build concern in the general public have garnered little attention; however, the producers of GM organisms have responded to their concerns and established extensive testing programs to be applied to each candidate organism that is produced. In the meantime, GM varieties of corn, cotton, soybean and rapeseed have been put into agricultural production and are now extensively planted. These crops, and the other, newer GM crops, have produced no problems and have pioneered a silent agricultural revolution in the USA.

Keywords Agriculture · Biotechnology · Genetic engineering · Genetically modified organisms · Public perception/opinion

Abbreviations

GM	Genetically modified
GE	Genetically engineered
DNA	Deoxyribonucleic acid
Bt	*Bacillus thuringiensis*
CDC	Centers for Disease Control and Prevention

1 The Beginning of Genetic Engineering

Genetics underwent a paradigm shift in the early 1970s. Genes available for the breeding of new varieties were no longer limited to the same genus and species. Instead, new technology allowed genes from any source to be introduced and incorporated into any genome. Termed lateral transfer, this technology is based on enzyme systems originally discovered in bacteria, and viruses that allowed genes to be isolated, prepared in quantity, and transferred and incorporated into the genome of any organism. The technology has been applied in the development of new strains in agriculture, science, and medicine, giving rise to what are called genetically modified (GM) or genetically engineered (GE) organisms. In this article we will deal with GM plants in commercial agriculture, focusing on their adoption and acceptance in North America.

By the 1990s the technologies for gene isolation and transfer developed with microorganisms [1] were being applied to higher organisms. In agriculture there was an obvious need for new methods to protect plants from foraging insects and competitive weeds. Genes with potential to deal with these problems had been identified in bacteria. The first of these genes encodes a protein that, when ingested, is toxic to insect larvae, but in a higher organism is digested without effect [2]. The second of these genes encodes a variant of an essential plant enzyme that makes the enzyme insensitive to the broad-spectrum herbicide glyphosate, commercially sold as Roundup [3]. These genes were taken from the bacterial genomes that harbored them, cultured in vitro, and introduced into commercial varieties of corn, cotton, soybean, and rapeseed, a highly divergent range of plant types [4]. Comparing selected GM isolates with their parent strain for vigor and crop yield showed no detectable differences, while biochemical analysis revealed expression of transferred genes and, more importantly, field tests demonstrated strong protection from insect pest and/or herbicide damage [2]. These positive results allowed imaginations to soar; the possibilities appeared boundless. Weed-free, pest-free, and chemical-pesticide-free plants, in addition to enhanced yield, enhanced nutrition, retarded spoilage, and a longer shelf life are just some of the properties that were predicted for GM crops.

2 Risks and Concerns

2.1 Multiple Gene Copies and Unanticipated Genetic Results

The first lateral gene transfers into plant cells were carried out using gene guns [5], a method considered crude in comparison to the biological vector-based transfers systems used today [6]. Gene guns literally shoot DNA into cells. The DNA goes everywhere and multiple copies of the transferred gene can be distributed across a recipient's genome. The ambiguity associated with this approach raised fears about the possibility of creating undetected, deleterious secondary effects [6, 7]. Additional concerns about GM strains focused on the escape of laterally transferred genes into other species. Specifically, Table 1 lists concerns about the adoption of lateral gene transfer technologies for the development of new plant varieties raised by US interest groups such as Greenpeace, Friends of the Earth, the Organic Consumers Association, the Union of Concerned Scientists, and the American Council on Consumer In-

Table 1 Risks and concerns about GM products

Food allergies
Food toxicity and anti-nutrients
Antibiotic resistance
Increased weediness
Horizontal gene transfer

Table 2 Common plant toxins and anti-nutrients

Toxin family	Examples of occurrence in plants	Effect on humans and animals
Cyanogenic glycosides	Sweet potatoes, stone fruits, lima beans	Gastrointestinal inflammation, inhibition of cellular respiration
Glucosinolates	Rape (canola), mustard, radish, cabbage, peanut, soybean, onion	Goiter, impaired metabolism, reduced iodine uptake, decreased protein digestion
Glycoalkaloids	Potato, tomato	Depressed nervous system, kidney inflammation, carcinogenic, birth defects, reduced iron uptake
Gossypol	Cottonseed	Reduced iron uptake, spermicidal, carcinogenic
Lectins	Most cereals, soybeans, other beans, potatoes	Intestinal inflammation, decreased nutrient uptake/absorption
Oxalate	Spinach, rhubarb, tomato	Reduces solubility of calcium, iron and zinc
Phenols	Most fruits and vegetables, cereals, soybean, potato, tea, and coffee	Destroys thiamine, raises cholesterol, estrogen-mimic
Coumarins	Celery, parsley, parsnips, figs	Light-activated carcinogens, skin irritation

terests. While all of the concerns listed are valid, and were especially so in the early years of this work, improved technologies in production of strains and more exhaustive characterization of product strains have considerably reduced risks. Genetic analysis of the recombinant strains reveals the number of gene copies inserted as well as the location of the inserted genes with respect to those of the parent genome. Such analyses are used to identify and eliminate recombinants with potential for alteration in expression of the parent organism's genes [7]. Additional tests on promising, transgenic isolates are performed to ensure that the expression of critical parent genes is unaltered. These tests include the determination of levels of production of indigenous plant toxins (listed in Table 2 [2]) and detection of production of proteins with allergenic potential [2, 5, 8, 9]. Acceptance of GM products has become a matter of trust that the work characterizing the recombinant isolates genetically and the associated comprehensive risk analyses testing have been responsibly performed and are reliable.

2.2 Escape of Transferred Genes

Another set of concerns regards the escape and dissemination of laterally transferred genes. The concern divides into two areas: the escape of genes that would make foreign or weed species more competitive (such as becoming resistant to glyphosate) and the transfer of bacterial antibiotic resistance genes (which are used as selective agents in lateral gene transfers) back into the bacterial kingdom and ultimately to pathogenic species. The first concern is a real possibility if there are genetically related species growing within pollen-transferable distance to GM crop agriculture. However, it can be argued that with the initial species used for GM agriculture (corn, cotton, soybean, and rapeseed) that in the USA and Canada there are no genetically compatible native plants [10]. Additionally, trans-genus gene transfer by natural mechanisms has never been observed, nor, in spite of centuries of trying, achieved. However, as GM crop development is expanded, a choice of culture habitat must be taken into account in applying the technology to additional species [9]. For example, the culture of a glyphosate-resistant commercial variety of bent grass in North America could open the way to the spread of the resistance gene into native, wild bent grass varieties. However, this should not be a concern for its use in lawns and golf courses in non-bent grass areas such as arid areas in the Middle East [11, 12].

The second concern about lateral transfer is that bacterial antibiotic resistance genes (used to aid the selective transfer of desired genes) cross kingdom lines and return to the bacterial world. Such transfer is an extremely remote possibility and would be an unprecedented event, not possible via documented, naturally occurring mechanisms of gene transfer [13] described below. Transmissible antibiotic resistance between bacteria was first encountered in the 1960s [14]. The resistance was shown to result from two factors: (i) a gene coding antibiotic resistance, and (ii) a DNA carrier molecule that is capable of being transferred from one cell to another. The rapid dissemination of antibiotic resistance in bacterial populations has been and continues to be a major concern in the practice of medicine. However, antibiotic resistance genes by themselves have proven to be a highly useful tool in genetic engineering. DNA that carries genes to be laterally transferred is joined to DNA carrying a gene that confers antibiotic resistance, and the hybrid DNA molecule is transferred to the recipient organism. Antibiotic-resistant clones are selected. Offspring that show antibiotic resistance are then tested for co-transfer of the gene of interest [2, 13]. The concern that GM foods developed in this way may become an additional source for the spread of antibiotic resistance to bacteria is without foundation because the second element, an appropriate DNA carrier molecule required for the transfer, is absent.

The transfer of DNA from one bacterial cell to another depends on conjugative plasmids, temperate phage or species specific transfer of naked DNA [15]. There is no evidence for conjugative plasmids or temperate phages that can cross between the plant and animal kingdoms, and there are only a few species of bacteria capable of taking up and incorporating naked DNA. In these species for the uptake of DNA to occur, the DNA must carry species-specific signature sequences that are rarely found in other DNAs [16]. It is very improbable that an antibiotic resistance gene present in a plant product without the appropriate signature sequence would find its way into a bacterium. This possibility is made more remote by the fact that there are enzymes in the digestive tract that would degrade the DNA [2].

Although there is an extremely low possibility for the transfer of antibiotic resistance genes from plants back to bacteria, the industry has not taken this possibility lightly and has adopted a precautionary approach in order to preclude any problems from this possibility. The genes that convey antibiotic resistance chosen for use in GM crop development provide protection against antibiotics that are clinically ineffective, either by their nature, or because resistance to them is already widespread [2, 13]. In the meantime, a search for alternative, less controversial, selectable markers has been initiated [17].

2.3 Creation of Superweeds

The concern that GM plants may escape to become weeds that cannot be controlled because they are resistant to herbicides is also unlikely. The majority of agricultural varieties are highly selected for yield of desired product and not for aggressive habitation. Most of them are poor competitors if left in the wild and are almost entirely dependent on humans for their perpetuation [18]. Plant breeders have selected for plants that have large, starchy seeds with thin coats that stay attached to the plant longer. They have also selected for plants that contain fewer seeds [2]. The factors that have been bred into plants to increase their value as a food source have ultimately curtailed their ability to competitively reproduce as a seed plant [18]. The genetic alterations that have been put into plants do not change this at all. The seeds are no more competitive or able to go out and establish themselves than they were before [19]. A greater risk may come from a broad application of glyphosate or another herbicide. The widespread and continued use of a single herbicide provides a real chance for the selection of spontaneous mutants in the background population of native species. Continued use of the chemical would only help the spread and eventual replacement of native flora, including weeds, by resistant variants [20]. The appearance in 2005 and spread of glyphosate-resistant southern pigweed (*Amaranthus palmeri*) in three areas of the USA (Georgia, Arkansas, and Tennessee) shows that these concerns are real and justified [21]. The same argument applies to the widespread use of Bt-protected crops. When insect pest variants that are resistant to the toxin appear, a continued cultivation of Bt crops would support replacement of the sensitive parent strain(s).

3 Case Studies: Starlink Corn and Monarch Butterfly

The general public's awareness of the potential problems arising from the employment of GM products was raised by two reports that were widely circulated in the press: wind dissemination of Bt pollen upsetting the food net outside of cultivated areas, and new allergens produced by Starlink GM corn. The report in the journal Nature states that there is a potential danger to the Monarch butterfly feeding on milkweed near fields growing corn producing Bt-resistant pollen. The study claims that the pollen, when ingested with the leaves, is poisonous to the insect larvae and could stunt its growth or kill it [22]. While subsequent studies showed that the larvae are susceptible to the Bt toxin, the likelihood of a sufficient amount of pollen falling on nearby leaves to cause larvae any harm is negligible [23–25], but the secondary reports gained only brief mention in the press [26].

Another problem widely reported in the press was of people reporting allergenic reactions following ingestion of foods containing Starlink corn. Starlink corn is a variety of GM corn that is Bt-resistant and resistant to glyfosinate herbicides. There was concern that Starlink corn might pose an allergy risk and accordingly it was not approved for human consumption; however, it was approved for use as animal feed. Because of miscommunication between farmers and Aventis (the company that produced Starlink corn) some of the GM corn was mixed in with conventional corn and used in food products highly consumed by humans, namely TacoBell and Mission foods. Twenty-eight cases of allergic reactions attributed to the ingestion of Starlink corn were reported [27]. The CDC conducted a thorough investigation of the individuals and concluded that the individuals did experience allergic reactions; however, there was no evidence that the reactions were associated with the presence of Starlink corn in the corn products they consumed [28]. Although these two reports of GM problems turned out not to be supported, the issue of potential problems and safety is really a case-by-case affair and concern about the technology remains unchanged.

4 Public Opinion and Perception

The negative publicity generated by the reports of pollen toxicity and Starlink corn allergens raised concerns about the public's acceptance of GM foods. Several surveys were carried out to gauge the public's attitude to GM products [29–31]. The results of the polls were inconsistent. Some surveys showed a 70% positive response when the questions were asked with reference to the benefits of GM products or a 75% negative response when questions were framed around the risks and concerns for the technology [32]. What these and other surveys showed is that the general public is ignorant of GM technology and its level of implementation in this country. The development of GM products is based on sophisticated genetic techniques, and the general public is woefully ignorant of even the basics of genetics. For example, a 2004 survey of a representative cross-section of the American public found that only 15% of the respondents were sure that the incorporation of a catfish gene into a tomato would not produce a fishy tasting fruit and, even worse, only 9% of the respondents were confident that tomatoes contained any genes at all [31].

5 The Present and Future Applications of GM Foods

GM foods were introduced commercially in 1994. Initially a limited number of farmers were willing to take the risk that GM crops would be marketable. Ultimately this concern turned out to be unfounded in the US market, as GM foods passed quietly into the food supply free from announcement or identifying labeling. As a result, the vast majority of people in the USA consume GM foods but are unaware that they do so; the products appear unchanged and no problems have been detected from their consumption. GM crops provide farmers higher yields and lower costs [33], leading to more farmers planting more acreage each year. Figure 1 [34] shows that for the last 7 years (1999–2005) the acreage of GM plantings in the USA and the world as a whole has been increasing exponentially. Additionally, the world, including the USA, shows

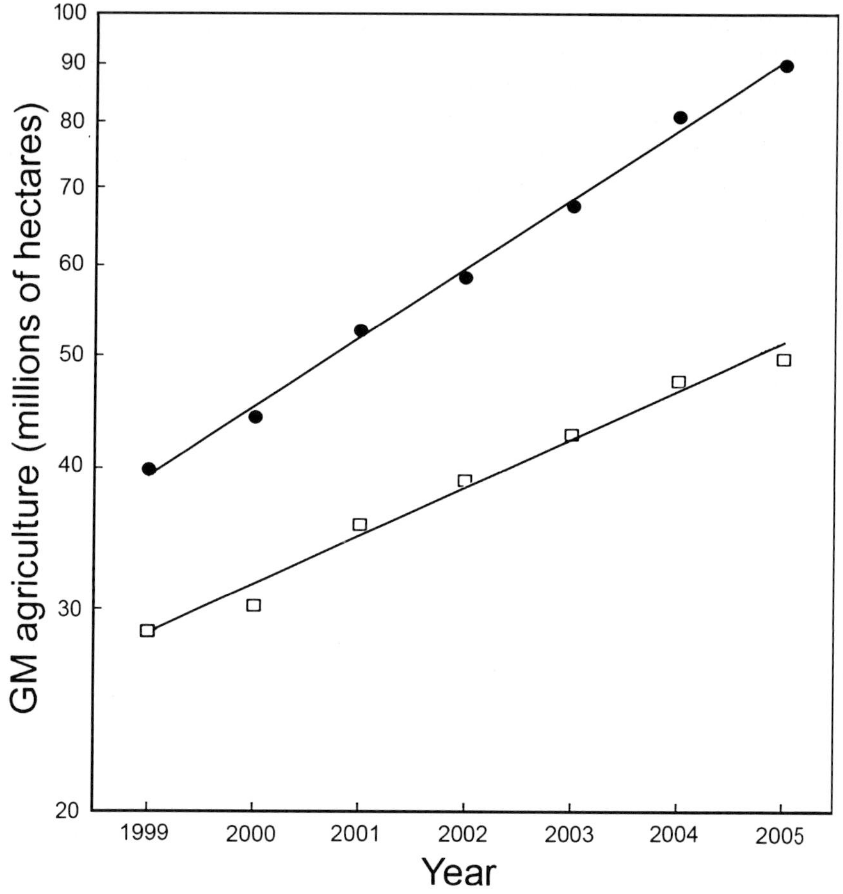

Fig. 1 Hectares used for GM agriculture in the world and in the USA. • Values for plantings in the world, □ values for plantings in the USA. The world value of 81×10^6 ha in 2005 is one third of the world's arable land under cultivation [34]

more than a doubling in the number of acres of GM crops planted over the period [34]. The USA had a head start on the employment of GM agriculture and, as of 2005, still accounted for nearly 55% of the world's GM acreage [34]. Genetically modified forms of three crop species, corn, cotton, and soybean are now well accepted. These three species account for 94% of the land cultivated with GM crops in the USA [34, 35]. A breakdown of the statistics for the planting of these crops in the USA [34] shows that by 2004, GM cotton had reached 76% replacement and its use was growing exponentially. Soybean and corn exhibit even greater rates of replacement, plantings extrapolate to 100% replacement by 2006 and 2008, respectively (data not shown).

The USA has, by default, become a giant demonstration project for the safety of GM crops. Still, when surveyed, consumers would like to know what foods contain GM products [30, 31]. However, with the depth of penetration of some GM products, it may be simpler to follow the lead of pesticide-free and organically grown foods and label products as non-GM or GM-free. At this time, however, the public's general lack of awareness or concern about GM foods leaves the food industry without any impetus to initiate labeling beyond that required by the Department of Agriculture for nutritionally altered GM products.

What is the future for GM foods? With any new product there are always risks, and unforeseen problems can arise. With GM foods, each new construct needs to be tested thoroughly to help ensure that the desired properties are the only changes to the organism [9]. However, the same argument applies for conventional breeding methods. For example, the use of conventional breeding methods to develop commercial varieties of potatoes and celery with high levels of insect resistance resulted in an increase of natural toxin production to levels that made the plants toxic to humans as well as to insects [2, 36, 37]. These results support the use of GM technology to incorporate a foreign gene with targeted toxicity as a less risky approach to restrict insect damage. Currently, two different Bt toxin genes have been introduced into cotton (Bollgard II from Monsanto) and corn (Herculex XTRA and YieldGard Plus from Pioneer) that together provide protection to rootstock as well as to top growth [38]. Monsanto is experimenting with Roundup Ready Flex varieties of cotton that contain two glyphosate resistance genes in order to raise glyphosate tolerance in mature plants [39].

Beyond crop protection there appear to be an unlimited number of possibilities for the incorporation of genes would add to the nutritional value of foods. One example is "golden rice". Much of Asia suffers from a diet poor in vitamin A. With rice, the primary food source in Asia, a project was initiated to construct a strain of rice rich in vitamin A. This goal was achieved by the incorporation of vitamin A-producing genes from the daffodil [40]. The resulting golden rice variety is now under cultivation to build sufficient seed stock, as it awaits full deregulation so that it can be put into general food production [41, 42].

Another example is high lysine corn. The lysine content of proteins found in corn kernels is inherently low, so low in fact that corn cannot be used as a sole protein source in the feeding of humans or livestock. With humans, corn is often served with beans to achieve a nutritionally balanced diet; with animals the amino acid lysine, produced from bacterial culture, is added to the feed to bring the level of lysine to an essential level. There was hope that conventional genetics could solve the problem, especially following the discovery of a naturally occurring mutant maize strain with kernels containing protein with a significantly raised level of lysine. However, in cultivation, the fragility of the kernels, which opens them to disease, and the still inadequate level of lysine to meet nutritional requirements doomed its general adoption [43].

Recently, a GM approach was used to resolve the lysine problem, which was to introduce a bacterial gene into the maize genome to raise the level of lysine. Two considerations were taken into account in the construction: first the location of the bacterial gene with respect to the maize genes and second, biochemical properties of the introduced lysine synthesizing enzyme. Knowledge of the maize genome permitted the placement of the bacterial gene under the control of a maize gene predominantly expressed in grain, and so did not alter the overall biochemistry of the plant. The bacterial gene for lysine production was chosen for its lack of biochemical regulation and the ability of its enzyme product to produce unrestricted amounts of lysine [44]. The resulting strain, termed Lysine Maize LY038, contains five times as much lysine in its kernels as standard corn, and three times as much as the high lysine mutant strain mentioned above [45], while neither altering the biochemistry in the rest of the plant nor compromising the quality of the kernels. In fact, the level of lysine is so high in this strain that it should be possible to mix it 1 : 1 with standard corn and still meet the dietary requirements of all animals.

A still greater possibility for GM technology lies in the expansion of the number and variety of plant species used in agricultural production. Today the vast majority of plant species are not acceptable for consumption as food because they produce and contain significant amounts of toxic compounds [46]. Eliminating toxin-producing genes could add valuable plant food resources and minimize the dangers of relying on too few species for our world food supply. The USA remains a verdant land for the continued development of GM technology.

References

1. Berg P, Singer MF (1995) Proc Natl Acad Sci USA 92:9011
2. Cornell University (2004) Genetically engineered organisms: public issues education project http://www.geo-pie.cornell.edu/gmo.html, last visited 11 Feb 2007
3. Daniell H, Datta R, Varma S, Gray S, Lee S (1998) Nat Biotechnol 16:345
4. Monsanto (2005) Monsanto Company, St. Louis, MO http://www.monsanto.com, last visited 19 Aug 2005
5. Henry SA (2002) Agricultural biotechnology: informing the dialogue. Cornell University, Geneva, NY, http://www.nysaes.cornell.edu/comm/gmo/PDF/GMO2002.pdf, last visited 11 Feb 2007
6. Nuffield Council on Bioethics (1999) Genetically modified crops. http://www.nuffield bioethics.org/go/ourwork/gmcrops/introduction, last visited 11 Feb 2007
7. Agbios (2005) Mon 810 Food safety assessment case study: Modification method. http://www.agbios.com/cstudies.php?book=ESA&ev=MON810&chapter=Modifcation, last visited 11 Feb 2007
8. Metcalfe DD (2003) Environ Health Perspect 111:1110
9. Stewart CN, Richards HA, Halfhill MD (2000) Biotechniques 29:844
10. Hails RS (2000) Tree 15:14

11. Welterlen M (2003) Engineered bents: a concern? Grounds Maintenance, Feb 2003, http://groundsmag.com/mag/grounds_maintenance_engineered_bents_concern/, last visited 11 Feb 2007
12. Lee L (1996) Plant Sci 115:1
13. Smalla K, Borin S, Heuer H, Gebhard F, Elsas JD, Nielsen K (2000) Horizontal transfer of antibiotic resistance genes from transgenic plants to bacteria: are there new data to fuel the debate? In: Proceedings 6th international symposium on the biosafety of genetically modified organisms, July 2000, Saskatoon, Canada. University Extension Press, Saskatchewan, pp 146–154
14. Kronenberger CB, Hoffman RE, Lezotte DC, Marine WM (1996) Emerg Infect Dis 2:121
15. Ochman H, Lawrence JG, Groisman EA (2000) Nature 405:299
16. Chen I, Dubnau D (2004) Nat Rev Microbiol 2:241
17. Ellborough K, Hanley Z (2001) Emerging plant biotechnologies: new ways to find needles in haystacks. Information Systems for Biotechnology, http://www.isb.vt.edu/news/2001/news01.Aug.html, last visited 11 Feb 2007
18. Janick J (1999) Plant Biotechnol 16:27
19. Crawley MJ, Brown SL, Hails RS, Kohns DD, Rees M (2001) Nature 409:682
20. Hartzier B, Owen M (2003) Status and concerns for glyphosate resistance. Integrated crop management. Iowa State University http://www.ipm.iastate.edu/ipm/icm/2003/3-17-2003/glyphosate.html, last visited 11 Feb 2007
21. Culpepper SA, Grey TL, Vencill WK, Kichler JM, Webster TM, Brown SM, York AC, Davis JW, Hanna WW (2006) Weed Sci 54:620
22. Losey JE, Rayor LS, Carter ME (1999) Nature 399:214
23. Hellmich R, Siegfried BD, Sears MK, Stanley-Horn DE, Daniels MJ, Mattila HR, Spencer T, Bidne KG, Lewis LC (2001) Proc Natl Acad Sci USA 98:11925
24. Minorsky PV (2001) Plant Physiol 127:709
25. Sears MK, Hellmich RL, Stanley-Horn DE, Oberhauser KS, Pleasants JM, Mattila HR, Siegfried BD, Dively GP (2001) Proc Natl Acad Sci USA 98:11937
26. Marks LA, Kalaitzandonakes N (2003) Agbioforum 5:43
27. Bucchini L, Goldman L (2002) Environ Health Perspect 110:5
28. Centers for Disease Control and Prevention (2001) Report to FDA investigation of human health effects associated with potential exposure to genetically modified corn: a report to the US Food and Drug Administration from the Centers for Disease Control and Prevention. Centers for Disease Control and Prevention, Atlanta, GA
29. Mellman Group (2003) The Pew initiative on food and biotechnology. Public sentiment about genetically modified food. http://pewagbiotech.org/research/gmfood/survey3-01.pdf, last visited at 11 Feb 2007
30. US Food and Drug Administration (2000) Report on consumer focus groups on biotechnology. http://www.cfsan.fda.gov/~comm/biorpt.html, last visited 11 Feb 2007
31. Hallman WK, Hebden WC, Cuite CL, Aquino HL, Lang JT (2004) Americans and GM food: knowledge, opinion, and interest in 2004, FPI publication number RR-1104-007. Food Policy Institute, Rutgers, NJ
32. Kolodinsky J, DeSisto TP, Narsana R (2002) Influences of question wording on levels of support for genetically modified organisms. Consumer Interests Annual, vol 48 http://www.consumerinterests.org/files/public/QuestionWording-02.pdf, last visited 11 Feb 2007
33. Marra MC, Pardey PG, Alston JM (2003) AgBioForum 5:43

34. James C (2004) ISAAA Briefs 34-2005: Global status of commercialized biotech/GM crops: 2005. http://www.isaaa.org/Resources/Publications/briefs/isaaa_briefs.htm, last visited 11 Feb 2007
35. Pew Initiative on Food and Biotechnology (2004) Genetically modified crops in the United States. University of Richmond, VA, http://pewagbiotech.org/resources/factsheets/display.php3?FactsheetID=2, last visited 11 Feb 2007
36. Zitnak A, Johnston GR (1970) Am Potato J 47:256
37. Berkley SF, Hightower AW, Beier RC, Fleming DW, Brokopp CD, Ivie G, Broame CV (1986) Ann Intern Med 105:351
38. Australian Pesticides and Veterinary Medicines Authority (2003) Public release summary on the evaluation of the new active *Bacillus thuringiensis* var. *kurstaki* endotoxins as produced by the *Cry1Ac* and *Cry2Ab* genes and their controlling sequences in the new product Bollgard II cotton event 15985. APVMA, Canberra, Australia
39. Bennett D (2005) A look at Roundup Ready Flex cotton. Delta Farm Press, 24 Feb 2005, http://deltafarmpress.com/news/050224-roundup-flex/, last visited 11 Feb 2007
40. Golden Rice Project (2005). Biofortified rice. Golden Rice Humanitarian Board, http://www.goldenrice.org, last visited 11 Feb 2007
41. Coghlan A (2005) New "golden rice" carries far more vitamin. New Scientist, 27 Mar 2005, http://www.newscientist.com/channel/health/dn7196–new-golden-rice-carries-far-more-vitamin.html, last visited 11 Feb 2007
42. Potrykus I (2005) GMO-technology and malnutrition, public sector responsibility and failure. Electronic J Biotechnol 8(3) http://www.ejbiotechnology.info/content/vol8/issue3/editorial.html, last visited 11 Feb 2007
43. Mertz ET, Bates LS, Nelson OE (1964) Science 145:279
44. Lucas DM (2004) Monsanto petition for determination of non-regulated status for Lysine Maize LY038. Monsanto Company, St. Louis, Missouri, http://www.aphis.usda.gov/brs/aphisdocs/04_22901p.pdf last visited 11 Feb 2007
45. Huang S, Kruger DE, Frizzi A, D'Ordine RL, Florida CA, Adams WR, Brown WE, Luethy MH (2005) Plant Biotechnol J 3:555
46. Ames BN, Gold LS (1990) Proc Natl Acad Sci USA 87:7782

Invited by: Professor Fiechter

Adv Biochem Engin/Biotechnol (2007) 107: 69–96
DOI 10.1007/10_2007_058

Published online: 31 March 2007

Agricultural Biotechnology and its Contribution to the Global Knowledge Economy

Philipp Aerni[1,2]

[1]World Trade Institute, Hallerstrasse 6, 3012 Bern, Switzerland
philipp.aerni@wti.org

[2]Center for Comparative and International Studies (CIS),
Swiss Federal Institute of Technology (ETH), ETH-Zentrum,
SEI F6, 8092 Zürich, Switzerland

Abstract The theory of neoclassical welfare economics largely shaped international and national agricultural policies during the Cold War period. It treated technology as an exogenous factor that could boost agricultural productivity but not necessarily sustainable agriculture. New growth theory, the economic theory of the new knowledge economy, treats technological change as endogenous and argues that intangible assets such as human capital and knowledge are the drivers of sustainable economic development. In this context, the combined use of agricultural biotechnology and information technology has a great potential, not just to boost economic growth but also to empower people in developing countries and improve the sustainable management of natural resources. This article outlines the major ideas behind new growth theory and explains why agricultural economists and agricultural policy-makers still tend to stick to old welfare economics. Finally, the article uses the case of the Cassava Biotechnology Network (CBN) to illustrate an example of how new growth theory can be applied in the fight against poverty. CBN is a successful interdisciplinary crop research network that makes use of the new knowledge economy to produce new goods that empower the poor and improve the productivity and nutritional quality of cassava. It shows that the potential benefits of agricultural biotechnology go far beyond the already known productivity increases and pesticide use reductions of existing GM crops.

Keywords Agricultural biotechnology · Cassava · Developing countries · Empowerment · Knowledge economy · New growth theory

1
New Growth Theory and the True Value of Technological Change

Most economists today are still trained in neoclassical economics. Its basic comparative-static assumptions of perfect competition, knowledge as a pure public good, and price-setting as market failure were very popular in the twentieth century because they enabled elegant formalizations of general and partial equilibrium models from the household economy to the world economy. This neoclassical approach is based on the assumption that all goods and technologies that could possibly exist, do already exist. This philosophy of plentitude [1] proves to be particularly inadequate in a knowledge economy where the exponential growth of knowledge leads to an exponential growth of the probability that new goods and technologies come into being and generate new markets. This process is not just the primary source of wealth and prosperity but also generates a social welfare surplus that cannot be captured by the innovating company itself. Paul Romer, the father of new growth theory, was able to highlight the social welfare impact of new goods within the neoclassical economic model [2]. In that article "new goods" primarily refer to capital goods that are used as an input for the improvement of the production of already existing commercial goods or the creation of entirely new goods (e.g., products derived from agricultural biotechnology are new goods that improve the quantity and quality of agricultural production).

1.1 Explaining the Knowledge Economy

New growth theory emerged in the 1990s in response to the inadequate assumptions of neoclassical theory. In his paper, "Endogenous technological change" Paul Romer (1990) showed that knowledge, unlike other production factors such as land, labor, and capital, is a non-rival good that can be used by many at the same time without loss in value. Thanks to the revolution in information technology, this knowledge can be reproduced at almost no additional costs. Yet, the creation of new knowledge is expensive since it requires large fixed costs spent on research and development (R&D). These costs also include the hiring of scarce and expensive human capital, the most sought-after resource in the knowledge economy [3]. It is therefore not surprising that those who create new knowledge want to make its use partially excludable through intellectual property rights (IPRs). This temporary monopoly right allows the owning company to extract a rent by putting the price of the new knowledge-intensive product above its marginal production costs. It thus enables the company to compensate for the high fixed costs that were spent on R&D, and, at the same time, provides incentives to invest again in improvement of the product.

1.2 Monopolistic Competition as the Primary Source of the Creation of New Goods

It is therefore monopolistic competition and not perfect competition (as portrayed in the idealized neoclassical theory) that creates new goods and new markets. The company that introduces a new good has the temporary power of setting the price of this good in the market (unlike companies in perfect competition, which are all assumed to be price-takers). Neoclassical welfare economists often denounce price-setting power as the extraction of an undeserved rent by a monopolist at the expense of the consumers, who suffer a deadweight loss due to the higher price they have to pay for the product. Even though empirical research by economists themselves showed that these deadweight losses are quite small [4], policy makers tended to identify monopolistic competition as market failure that needs to be addressed by government intervention. This thinking is, however, based on two contested assumptions: (i) knowledge is a non-excludable public good funded by governments and produced at public universities and national research institutes and (ii) monopolies exist primarily because they repress competition through high barriers to market entry generated through collusion and political lobbying.

These two assumptions are not wrong but they are not the whole truth. Governments indeed fund the production of knowledge and make sure that it is widely accessible. But, only the private sector can convert new knowledge into successful markets for new goods, technologies and services, and this conversion also requires a fair amount of investment in R&D. Moreover, the decision to invest in a new technology is particularly risky and expensive due to high R&D costs and a high degree of unpredictability with respect to the estimated demand, the length of the approval process, social acceptance, and potential liability costs. Therefore, a company will only invest in a new technology if there is a prospect of making a considerable profit. Such a profit is possible if a company can obtain a temporary monopoly right through the protection of its intellectual property (the temporary right to exclude others from copying and selling the same product). This makes the non-rival good partially excludable. Yet, such a monopoly is not just extracting a rent from consumers through a higher price but also creates a new good that produces a social welfare surplus, which cannot be captured by the company itself (e.g., more employment, more tax revenues, more knowledge in the public realm through patent disclosure, economic spillovers leading to generic products that are also affordable to poorer consumers/producers, etc). Thus the more useful knowledge that is generated in society the higher the likelihood that new goods will be created. Therefore, this new knowledge economy is no longer about substituting one existing good for another existing good but about the constant introduction of new goods.

1.3 Social Welfare Generated by New Goods

The social welfare that results from the introduction of a new good is not new but was already noted by a French engineer called Dupuit in the nineteenth century [2]. He calculated the cost of building a bridge and the minimal toll the users of the bridge need to pay to reimburse the fixed costs for building the bridge. He was able to show that the entrepreneur who builds the bridge is constrained in his efforts to extract a maximal rent from the users, because if the toll is too high the user might simply not use the bridge (assuming that the users are acting in a competitive world with scarce resources themselves). He therefore concluded that the entrepreneur can never capture all the benefits of building the new good "bridge". The same is true for a company that wants to develop a new technology. Instead of extracting an additional rent through a toll (as in the case of a physical good), it would do it through a royalty fee on the patented technology.

1.4 Adding a Dynamic Dimension to Welfare Economics

The creation of new goods that emerge from monopolistic competition can be illustrated by means of simplified version of a partial equilibrium model

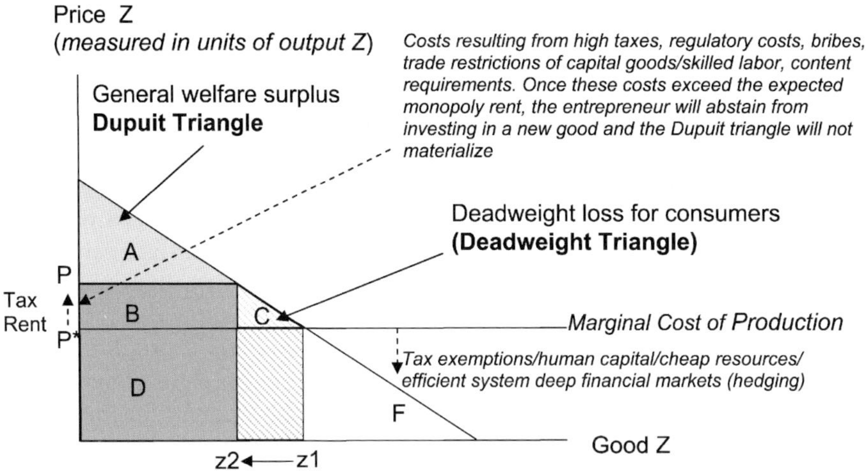

Fig. 1 The welfare surplus generated by the introduction of a new good, according to Romer [2]

adopted from Romer [2] (see Fig. 1). It represents a partial equilibrium model with the x-axis referring to the amount of production of good Z and the y-axis the price per unit charged by the company. The marginal cost of production indicates any additional cost required to produce a next unit. This marginal cost curve is flat rather than increasing because it only represents the variable costs of production (below the line) that are assumed to remain constant with increasing production in view of the low and relatively stable reproduction costs of an innovation. The price is higher than the marginal production cost because the company aims at reimbursing the high fixed costs spent on the development of good Z (not represented in the marginal cost of production) and making profits that allow for further investment in R&D.

Neoclassical economists interpret this graph as a typical case of a market that is dominated by a monopolist: there is only one producer of good Z, who has the power to determine the scale of production and set the product price in a way that maximizes the expected returns (the sloping demand curve illustrates how the price increases with decreasing output). In order to illustrate the monopolist rent, the neoclassical economist would point at rectangle B, which represents the surplus the monopolist extracts through market power, and also triangle C, which represents the deadweight loss for consumers who have to pay a higher price for the good. However, if this monopolist has obtained his position not through collusion but the investment in the development of a new good, an additional social welfare surplus, triangle A, comes into being. This triangle has been ignored by economists prior to Romer. It represents the social welfare surplus that is generated through the new good. The monopolist cannot capture this triangle because if, as Dupuit already recognized, he raises the price above this level, he would lose the market (the price would be too high for potential buyers that are constrained by their own competitive markets).

Romer concludes that if economists became aware of this triangle A (which he also calls the Dupuit triangle), they would realize that the primary source of wealth and well-being in society is not based on perfect competition but on the introduction of new goods and technologies. New technologies may create new inequalities and risks at the beginning, when only few have access to the relatively expensive technological innovation and accidents may happen due to lack of experience with the new technology. But in the long run many new competitors enter into the market, increase the total offer, constantly improve the safety of the technology, and lower the price. Eventually, the product becomes cheaper and safer and thus generates a global mass market. At this stage, it could also become a tool of empowerment for the people who previously lacked access to the technology [5].

1.5
Role of Multinationals in Economic Growth

Currently, the main players in the global mass market are multinational companies. They are interested in selling their goods in all parts of the world. It is true that their product must be sold at a much lower price in developing countries with low purchasing power, but the fact that they have price-setting power allows them to exert price discrimination. They can charge higher prices in developed countries and lower prices in developing countries. Moreover, poor developing countries may be high-risk markets but they generally offer greater long-term growth potential. There are certainly multinationals that tend to abuse the system by charging prohibitive prices in certain countries, aggressively push the strict enforcement of their patents and later on their extension, lobby for permissive regulation, and use the returns to enrich themselves rather than invest them in new R&D. However, it can be assumed that those companies may eventually lose out in competition because they had spent their scarce financial resources on preserving short-term political and market power instead of investing them in R&D to ensures the long-term survival of the company in a competitive and innovation-driven market.

1.6
Responsibility of National Governments

Whether a multinational company decides not just to sell but also to develop and produce a new good in a particular country, depends to a great extent on national government policies. National governments can discourage such investment by imposing high profit taxes, trade restrictions on essential capital goods, prohibitive safety regulation, and inefficient and burdensome government bureaucracies. Corruption and weak property rights can additionally increase the costs for the company until the point is reached where a company decides not to invest anymore in a developing country, despite cheap labor and abundant natural resources, because the increasing costs exceed the additional revenues expected through the higher price they intend to charge (see Fig. 1). So the good will simply not be produced, which means that the respective country loses the social welfare benefits of the Dupuit triangle.

Instead of just taxing and regulating companies and pushing up the bar of their marginal costs of production, governments can also serve as facilitators and encourage investment through tax exemptions for R&D, political stability, a valuable stock of domestic human capital, dependable public infrastructure, and a relatively open and well-developed financial market (that improves the possibility of hedging risks).

In this context, governments in developed countries have many means available to facilitate private sector investment. This allows them on the other hand to increase the cost of regulation of a company (e.g., by pushing up environmental and food safety standards). Poor countries, however, are in a different position: because the market is tiny and the state budget too small to improve investment conditions, additional regulations quickly erase the profits that foreign companies can expect. This again leads to the loss of the Dupuit triangles, which are much larger than the deadweight loss triangles for consumers (triangle C). Paul Romer [2] therefore concludes that "badly designed policy interventions do not come from their effects on the static allocation of resources between the activities in an economy that already exists. Rather, they come from the stifling effect that the distortions have on the adoption of new technologies, the provision of new types of services, the exploitation of new productive activities, and on imports of new types of capital goods and produced inputs".

2
How to Use Knowledge and Technology for Development?

As illustrated above, the primary contribution of companies to general social welfare may not occur through general taxes (as widely assumed) but through the generation of new goods and services. Yet, the problem is how to get the private sector to produce valuable goods for markets that are too small to be worth investing in? Often these markets are not served because of generally low purchasing power or small market size.

This is a serious problem because the Dupuit welfare triangles for such goods would be huge. For example, orphan crops with higher productivity and enriched nutritional quality (e.g., protein-enriched cassava) or orphan drugs against communicable diseases (e.g., Malaria vaccine) have the potential to improve public health in developing countries enormously. Old welfare economics would suggest that the insufficient production of such goods is a typical example of market failure.

New growth theory, in turn, would argue that it is state failure because governments fail to design appropriate incentives for the private sector to collaborate with universities in efforts to create new goods or improve existing goods that are of high value in the fight against poverty and for improvement of the natural environment.

Since neoclassical welfare economics shaped agricultural and development policies in the twentieth century and continues to do so, it is important to first understand the achievements, but also the unintended side effects, of this comparative–static approach during the Cold War.

3
Cold War Economics and its Impact on Agricultural Policy and Research

The rise of neoclassical welfare economics began after World War II. At that time, Western policy decision-makers were not just concerned about economic growth (which was strong anyway due to the reconstruction efforts) but also about the ideological mindset of its citizens. It was assumed that economic development may result in increased social inequality and that this might cause public disillusion with capitalism and a longing for communism. Welfare economics was regarded as the ideal instrument to identify the aggregated preferences of a society (in terms of expected social outcomes) and to design policies that addressed such preferences (e.g., minimal social inequality). These aggregated preferences were portrayed in the form of a normative social welfare function. Subsequently, positive analysis was applied to maximize allocative efficiency of the public policy measures designed to achieve the resulting social objectives. The celebration of the elegant formal language of general and partial equilibrium models in welfare economics and the belief that the public sector must assume the role of a rational social planner that addresses the problem of market failure influenced many areas of public policy.

Yet, the great welfare economists at that time Paul Samuleson, Kenneth Arrow, and Robert Solow were too quick in identifying market failure when it came to the management of public goods. Samuleson [6] used the example of the lighthouse to illustrate the nature of a public good. He argued that since none of the shippers would be willing to build such a lighthouse (in view of the others who would just free-ride) it must be a public good that is based on non-rivalry and non-excludability and therefore should be provided by the public sector. Ronald Coase showed that this argument does not correspond to historical facts: lighthouses that were financed by user fees paid by the shippers existed in Europe in the eighteenth century, as he pointed out in his 1979 publication [7].

Kenneth Arrow [8] and other welfare economists portrayed the state as a rational social planner that looks at aggregate social preferences and, accordingly, allocates the scarce public resources in a pareto-optimal way (making at least someone better off without making anyone worse off). Buchanan and Tullock [9] discovered the flaws in such assumptions by highlighting the fact that the democratic decision-making process is not a rational process on an aggregated level as Arrow assumes. Political actors pursue their own self-interest and are not driven by the desire to maximize social welfare. Moreover, there is no such thing as a rational social planner unless it refers to an all-knowing dictator.

Finally, Robert Solow [10] managed to combine conventional growth theory with neoclassical welfare economics. He argued that knowledge must be funded by the public sector because it is a public good (assuming that it is non-rival and non-excludable). As a result, technological change was treated as an exogenous factor that can be perfectly integrated into the neoclassical model of perfect competition, where companies are portrayed as passive price-takers. Romer finally challenged Solow's growth model with his paper on endogenous technological change [11]. He showed that it is not perfect competition but monopolistic competition that generates new goods and services. Moreover, his model was able show that companies also have to invest in R&D (something the Solow model did not account for. A more detailed discussion of the theoretical aspects of new growth theory and the knowledge economy has been published by Aerni in the ATDF Journal [12]).

4
Entrenched Interests in the Post-Cold War Economic Community

In view of all these developments, one would therefore assume that economics has evolved by accepting this shift in paradigm. However, textbook economics continues to be welfare economics and technology is still largely treated as exogenous. Those economists who deal with issues such as technological innovation and monopolistic competition are generally advised to leave the economics departments and to join business schools [13]. This ongoing reluctance to embrace the new paradigm may also be related to the entrenched interests within the academic community of economists. It highlights once again the contradiction that is so obvious in social science. On the one hand, most social scientists claim to respect Popper's concept of falsification, but then who is really eager to falsify the theory that he or she helped create? Unlike in the natural sciences, where an increasing gap between theory and experimental outcome leads eventually to an adjustment of the theory, there is no such pressure in the social sciences. Most theories that history has proved wrong or inadequate are still taught at universities and continue to expand in the form of insider research communities. These research communities have their own peer review process, which relies on reviewers from the inner circle of professors that are in charge of maintaining the dogmas of the theory and have an interest in keeping the theory alive [14].

4.1 The Mindset of Agricultural Economists and How it Influenced Agricultural Policy and the Green Revolution

The first wave of globalization in the second half of the nineteenth century led to a rapid decrease of transportation costs and erased many of the geographical barriers that previously protected local agriculture from foreign competition. This threatened the livelihoods of many farmers in the early stage of industrialization and governments started to be concerned about the capability of their respective countries to ensure the national food supply in times of war [15]. Political and economic (later also social and environmental) instabilities were identified as negative external effects that need to be addressed by the public sector [16]. At the outset, state intervention was mainly justified in the name of managing the public good called "national food security". Agricultural economists were hired as social planners to ensure the effective management of this public good. The planning models they used at a later stage (e.g., linear programming) to calculate how certain normatively set policy objectives can be achieved most effectively, were largely developed by scientists in the former Soviet Union. Even though these policy planners quickly realized that a democracy is not about the joint effort to design a rational policy but about the self-interested search for access to scarce public resources, it did not hinder them from sticking to the principles of welfare economics and ignoring the role of political economy.

4.2 Agricultural Policy During the Cold War

Agricultural economics embraced the theoretical concept of the so-called agricultural treadmill developed by Cochrane [17]. In this concept, farmers produce a homogenous (presuming that there are no differences in quality) and inferior (because of its low income elasticity: the higher the income the lower the share of the household budget for food) commodity in the form of food. They are portrayed as passive price-takers in a market of perfect competition. The role of technology is reduced to its potential to increase agricultural productivity (while its potential to improve food quality is not addressed). Since farmers are standing in perfect competition they are assumed to produce at the level where their marginal costs just equal their marginal revenues. According to the concept of the agricultural treadmill, it is possible that certain farmers adopt a new technology that allows them to lower their production costs and produce more efficiently. This gives them a temporal advantage and thus a windfall profit. Yet, this advantage is quickly erased because all the competitors will have to follow suit if they want to stay in business.

The agricultural treadmill is portrayed as the main reason for the surplus in food production and the decrease in relative food prices. Unsurprisingly, agricultural economists concluded that the treadmill, and with it technological innovation, largely benefits food consumers at the expense of food producers. They argued that the agricultural treadmill produces a sort of market failure since farmers get poorer even though they produce more, due to the inferior prices. Moreover, the resulting intensification of agriculture will destroy the environment and family farming and negatively affect the quality of food. Yet, in retrospect, even agricultural economists would admit that it was probably not the agricultural treadmill, but the market-distorting instruments of agricultural policies that provided the biggest incentives to adopt monoculture practices and degrade the environment and food quality. One only needs to go and watch the movie *We feed the world*, produced by Erwin Wagenhofer in 2005 (the most successful Austrian documentary movie ever) to get a picture of the unappealing endless number of greenhouses in southern Spain that focus almost exclusively on intensive tomato production. Erwin Wagenhofer, who is an urban dweller with little knowledge of agricultural policy, blames the corporate world for all this misery. Yet, in fact intensive tomato production in Spain is a result of EU subsidies. The same goes for olive tree monoculture in Spain and Greece, overfishing in the Atlantic Ocean, excessive growing of low-quality wine in France, and many other subsidized products. All these practices are not just harming the environment but they also discourage innovation and tend to make food quality worse – for why should these producers care about innovation or pleasing consumer taste if the money comes from Brussels anyway?

4.3 Agricultural Policy After the Cold War

In the 1990s, agricultural economists admitted that certain policies produced "suboptimal" results despite the rational social planning. They recommended a switch from production-tied subsidies to income-support subsidies. The new objective was the maintenance of a strong, healthy, and environmentally sustainable agricultural sector. As a consequence, things like agrobiodiversity, food safety, decentralized settlement, and custodianship to cultural landscapes were declared to be the new public goods that are provided by farmers – after the old public good of maintaining food security became somewhat obsolete in view of the production surplus and the end of the Cold War. It provided the best justification to keep agricultural economists employed as social planners and continued to use all the old planning models that are focused on creating optimal allocative distribution in areas where the market presumably fails to do so. But did the market really fail or are these bureaucrats increasingly managing state failure?

There is increasing evidence that the new agricultural policies and the new justifications for government intervention in agriculture did not bring the expected improvements: Direct payments were designed to mitigate the structural change that was expected to result from slightly more open agricultural markets, as demanded by the WTO Agreement on Agriculture (AoA). Yet, direct payments proved to be an obstacle to structural change because they increased the value of land and discouraged many farmers from becoming more innovative and competitive. At the same time, the new normative goals of agricultural policy to promote environmental, social, and economic sustainability through compliance schemes (e.g., agro-environmental measures/labeling schemes in return for more direct payments and premium prices) once again turned out to be in the best tradition of government-sponsored agricultural economic research, namely suboptimal. Environmental improvements were relatively meagre and largely achieved through more efficient input technologies. Moreover, a large evaluation of agro-environment measures in Europe showed that such measures hardly ever contribute to an increase in valuable biodiversity [18]. There also seems to be a correlation between the amount of direct payments a rural region receives and its economic decline [19]. This is not surprising in view of the fact that a high dependence on direct payments is not an attractive way of life for the young people so they look for opportunities elsewhere to participate in the new knowledge economy. Apart from this, the private sector is reluctant to invest in heavily subsidized regions because of the receiver mentality of the people and high production costs (pushed up indirectly through direct payments).

In spite of the timid opening of agricultural markets, agricultural trade has hardly increased over the past two decades. One major reason for that is the WTO AoA itself. It is primarily focused on a gradual improvement of market access rather than the reduction of domestic support measures. But, ultimately, it is domestic support measures that result in wrong market signals, overproduction, and subsequent market access restrictions [20]. The fatal consequence was that the amount of domestic support did not decrease but was just moved from so-called "actionable" subsidies (placed in the amber box of the AoA to "non-actionable" subsidies (placed in the blue and green box of the AoA). In WTO terminology, subsidies are identified by "boxes" which are given the colors of traffic lights: green (permitted), amber (slow down – i.e., to be reduced), red (forbidden). Export subsidies are prohibited and therefore fall into the red box. This is, however, not strictly applied in agriculture where export subsidizes are still tolerated. The amber box contains all trade-distorting domestic support measures. Any support that would normally be in the amber box, is placed in the blue box if the support also requires farmers to limit production. Green box subsidies are meant to be non-trade-distorting.

It was assumed that non-actionable subsidies would not be trade-distorting but it turns out that they are. At any rate, this shift kept social planners employed and did not force anyone to look at theory. But are these policies sustainable and do they really benefit farmers? In consideration of what we now know, the answer is unlikely to be yes. A parallel development with a similar ambiguous outcome happened in the international arena where the primary concern was to develop new agricultural technologies to enable developing countries to become self-sufficient in food production.

4.4
Theoretical Thinking Behind the Green Revolution

The same agricultural economists who argued in the 1940s that the Western states needed to maintain a healthy agricultural sector through subsidies and border protection argued that developing countries must be assisted in the development of new varieties and modern irrigation systems in order to boost food production and avoid hunger and starvation. It was assumed that the private sector would have no interest in investing in technologies that would serve poor farmers in developing countries. Therefore, public investment in international agricultural R&D was declared to be a public good that must be managed by the public sector (following the Solow model). The resulting global public sector initiative is today widely known as the Green Revolution. It was to a large extent a US-driven effort to improve food security in the non-aligned developing world as part of a global containment strategy against communism [21]. The United States Agency for International Development (USAID) and the Rockefeller Foundation were the main financial contributors to the establishment of the first Centers of the Consultative Group of International Agricultural Research (CGIAR) in developing countries. These CGIAR centers enabled Western scientists to work in well-equipped research centers in developing countries and design high-yielding varieties of major food crops such as maize, wheat, and rice. The new varieties were subsequently distributed in rural areas through government agencies. The private sector was hardly involved, even though it later benefited from the scientific knowledge generated through this international undertaking. The research at these centers (CGIARs) contributed to significant increases in agricultural productivity and technology transfer to local universities and national research institutes in developing countries.

4.5
International Agricultural Research During the Cold War

There is no doubt that the Green Revolution greatly contributed to global food security through the excellent international agricultural research that was conducted at CGIAR centers during the Cold War. However, the interaction between Western scientists, who developed high yielding varieties, and local farmers in developing countries who adopted these varieties through the national seed distribution programs, was rather poor. This led to some long-term problems such as inadequate use of pesticides, insufficient operation and maintenance of irrigation systems, little seed choice for farmers, and monoculture practices [22]. In addition, farmers in marginal regions did not benefit to the same extent from these new hybrid varieties that were mainly designed for favorable agricultural conditions with access to fertile soil, irrigation, markets, and essential inputs [23].

Left-wing development activists point at these unintended side effects of the Green Revolution and denounce them as the destructive forces of science and business, and they conclude that environmentally destructive monoculture practices must be part of the capitalist logic. Yet, as highlighted in the case of the documentary of Erwin Wagenhofer, these undesirable side-effects are a result of too little rather than too much private sector involvement. For example, public sector researchers based at CGIARs did not have to bother much about the real and complex set of problems that farmer face in the field or the particular consumer taste of different cultures. They could just focus on plant variety traits that would increase yields and then select the elite varieties and hand them over to national agencies for distribution. As a result, the private sector may have had little interest in investing in the development and commercialization of new varieties in developing countries and in competing with the public sector, which would distribute the seeds almost for free. Thus, the private sector largely stayed out of the Green Revolution. This explains, for example, why the greatest bottleneck in many developing countries is probably the absence of a local seed industry. It also explains why many Filipino consumers prefer to buy rice from Thailand, which is the greatest exporter of high-quality Indica Rice, but actually never embraced the large-scale adoption of high-yielding rice varieties. They say it simply tastes better than the rice varieties that were bred by the International Rice Research Institute (IRRI) and widely adopted by Filipino farmers [22].

4.6 International Agricultural Research After the Cold War

After the end of the Cold War, foreign aid was cut in almost all state budgets of developed countries [24] and public sector funding for international agricultural research decreased significantly. Right-wing politicians argued that there was no need for further investment in CGIAR research because the Green Revolution has already largely achieved its purpose of eliminating hunger. This argument is quite cynical considering the fact that there are still over 800 million people worldwide suffering from hunger and malnutrition. Left-wing politicians, in turn, were using the familiar but flawed argument that there is enough food around but that it is badly distributed. This led them to the conclusion that there is no further need for investment in technology but instead just a need to bring the food to the poor. Agricultural ministries in developed countries that still do not know how to get rid of production surpluses would most certainly welcome the idea. Yet, the fatal consequences of such dumping policies are widely known: local farmers in developing countries that cannot compete with donated food are forced to abandon farming because of lack of revenues. Thus, such policies are likely to worsen food self-sufficiency and increase dependence on Western food aid.

Even though the "distribution problem" argument is still widely used by teachers in high-schools, it is rejected even by left-leaning development activists who now embrace the paradigm that farmers in developing countries need to be assisted in growing their own food in a sustainable way. Yet, the problem with Western non-governmental organizations (NGOs) that pursue this approach in developing countries is that they generally dismiss the role that business and new technologies have in agricultural development, arguing that it would introduce a capitalist logic that is not compatible with local traditions. They believe that farmers should rely on their traditional low-input and low-tech practices. They may assist them in finding slightly better techniques of soil fertility and integrated pest management but, in general, farmers are encouraged to use the agricultural practices they would use anyway. Subsequently, these Western NGOs help them to export the harvested agricultural products to developed countries where they are sold under different kinds of environmental and social labeling schemes. Such a strategy resembles the top–down approach of the Green Revolution.

Both strategies assume that there is a sort of market failure because business does not care about the poor. This would produce negative externalities such as increasing social inequality, hunger, and malnutrition that must be addressed by responsible Westerners. The only difference is that one approach looks at modern technology as the solution whereas the other approach sees it as the main problem. However, the ideological mindset of anti-technology NGOs is likely to harm poor farmers in developing countries more than the previous overemphasis on public sector R&D. Farmers need to become actively involved in the process of technological change and they need to learn how to take advantage of the emerging knowledge economy. This will eventually lead to more self-confidence and entrepreneurship and result in increases in agricultural productivity and nutritional quality of the traditional food crops. This is especially true for Africa, which did not benefit from the first Green Revolution.

It is often argued that local entrepreneurship in developing countries is hampered by the absence of land titles in the informal sector. Yet it is still unclear whether the assignment of land titles effectively results in more entrepreneurial activity independently of the other local circumstances. More important is the creation of an entrepreneurial infrastructure that lowers local market transaction costs and opens access to new markets. Local entrepreneurs can further be encouraged to become low-tech innovators through the establishment of petty patent systems [25].

In 2001, the UNDP Human Development Report was titled *Making new technologies work for human development* [26]. It attempted to counteract the misconception of the supposedly negative role of technology and the private sector in sustainable development and was promptly attacked by sustainable development activists. This is a pity because this report merely reminded policy-decision makers that there is Principle 12 in the UNCED Rio Declaration, which emphasizes the important role of new technologies in sustainable development.

It seems that neither agricultural economists that helped shape the Green Revolution, nor Western NGO leaders that advocate participatory approaches in agricultural development can see any benefit in getting the private sector more involved in agricultural development and encourage local entrepreneurship. This may be related to the fact that they tend to use theoretical concepts that might have looked reasonable in the Cold War economy, but they are rather outdated in the new knowledge economy.

5 The New Knowledge Economy and the New Rules of the Game

The two major driving forces of the new knowledge economy are the revolutions in information technology and biotechnology that took off in the 1970s and 1980s. Both revolutions started initially at universities and were strongly supported by the public sector. However, when the first prototypes of commercial interest emerged, the university-based inventors decided to seek intellectual property protection for their inventions in order to set up their own businesses in the form of spin-off firms. Some of them eventually established highly successful multinational companies themselves, others were able to partner with multinationals in the commercialization of the technology, others focused on licensing out their patented technology to whoever was interested in using it, and others simply lost out to entrepreneurial young outsiders that quickly grasped the economic potential of certain clumsy prototypes and improved them to a level where they could become commercial successes [27].

Both, the IT industry and the biotechnology industry have matured over the past decade. As a consequence, the costs of IT and biotechnology products and tools have decreased significantly and are now reaching a far wider customer base. Unlike in the old economy where most developing countries merely played the role of suppliers of primary commodities and lacked the critical base of domestic human capital to make use of modern technology to develop their homegrown technologies, the new knowledge economy allows them to participate in the global economy to a far greater extent.

5.1
Effects of Information and Communication Technologies

Thanks to all the new communication and information technologies, new knowledge spreads more quickly and widely, international research networks become much more extensive and effective, outsourcing business activities from simple accounting to R&D has become an integral part of the strategies of multinational companies, and global venture capital is increasingly invested in talented techno-entrepreneurs in developing countries. The resulting rise of many developing countries in science, culture, business, and political power makes the jargon of the North–South dialogue of many Western development activists look increasingly old-fashioned. South–South business investments and research collaborations are growing five times as fast as their North–South equivalents [28]. Moreover, many big companies in the South are starting to even gobble up companies in Europe and the USA.

5.2
Effects of the New Tools of Biotechnology

There is a widespread prediction that the biotechnology revolution, powered also by the information technology revolution, will eventually transform a rather dirty agrochemical and petrochemical industry into a cleaner biology industry [29]. The potential economic, social, and environmental welfare benefits of this transformation are enormous, and this time it is likely that developing countries with a critical domestic knowledge base will be at the forefront of new product developments. If mankind is serious about protecting the natural environment and ensuring access to food, the growing demand for food over the next 50 years should not be met by further colonizing of pristine ecosystems but rather by raising productivity on existing farmland. Agricultural biotechnology is not just ideally positioned to meet this challenge but also likely to produce new food products that are safer, more nutritious, and tastier. The potential environmental and health risks of biotechnology must be taken seriously, but after 10 years of experience and innumerable public risk assessment studies there is no indication that existing GM crops pose any risks that go beyond the risks known from conventional crops. Moreover, the ethical concerns raised about the current techniques of genetic engineering could quickly be overturned by the emergence of completely new transformation techniques and the advancements in genomic research. But, one ethical concern will certainly not go away and that is the crucial aspect of social equity.

The private-sector-driven biotechnology revolution may result in enormous social inequalities because the least developed countries that have simply no means, no critical knowledge base, and no attractive markets to participate in this emerging sector may once again be left out. As a result, the new products would merely improve the needs of affluent societies that can promise a high return on investment while the basic needs of the poor remain unaddressed.

5.3 Management of Public Goods in the New Knowledge Economy

As mentioned earlier, an exponential increase in knowledge leads to an exponential increase of the probability that new products and services will come into being. These new goods and services generate innumerable new Dupuit welfare triangles - but only for those societies that do not discourage their introduction and those that have sufficient purchasing power and market size to attract them. Therefore, there is a high likelihood that the knowledge economy will increase global inequality unless national governments and international organizations design policies that ensure that the new technologies will also benefit and eventually empower people in the least developed countries.

However, it would be a mistake to address the challenge by simply embracing a second Green Revolution because, as explained above, the underlying principles of welfare economics are no longer applicable to the rules of the new knowledge economy.

The belief that public goods should be provided exclusively by the public sector ignores the fact that the private sector increasingly contributes to the production of public goods (e.g., clean technologies, more efficient use of natural resources) [30]. The public sector should therefore not assume tasks that the private sector might provide in a more efficient way and in better quality (more focused on consumer/client needs) but should learn how to better play the role of a facilitator of private sector activities that generate large Dupuit welfare triangles. As shown in Sect. 1, the generation of these social welfare triangles requires high fixed costs that are spent on large investment in R&D, physical infrastructure, and product development. Companies are often unwilling to invest such high fixed costs in the development of a new good unless the resulting market is expected to be profitable. This also explains why the first prototypes of new technologies were almost always designed in university rather than in corporate laboratories [31].

Throughout the history of technology we can always observe the same pattern: there is the curiosity-driven researcher funded by the public sector who has no immediate interest in business. But, there is also the bold entrepreneur who uses the knowledge generated by the curiosity-driven researcher to design a new prototype that is patentable. The prototype is subsequently adjusted to market needs, and finally commercialized on a large scale. Both characters, the inventive researcher and the entrepreneur, are needed to create social welfare triangles. Sometimes the curiosity-driven researcher and the profit-oriented entrepreneur can be one and the same person. But often the inventor is not necessarily a good entrepreneur and the good entrepreneur is not necessarily good at inventing. Anyway, without the entrepreneur who is primarily focused on creating a new market and making large profits as a temporary monopolist, the fruits of science would never become real.

This is one of the most important insights, and it was ignored during the Green Revolution when research, product development, and distribution of new agricultural technologies was entirely managed by the public sector. The private sector is simply better at converting knowledge into useful goods and services. The public sector should mobilize this strength of the private sector for the management of public goods rather than try to duplicate it.

Today it is important that the public sector first identifies a list of biotechnology products that could potentially generate large social and environmental welfare benefits (e.g., biofuel generated from cellulose, drugs for diseases that are extremely rare, vaccines that protect people from sudden outbreaks of communicable diseases). It should then offer university research teams funding to develop a first prototype of such a new product. At the same time, it could offer a generous award to the research team that first develops a dependable prototype that is sufficiently attractive to be licensed out to the private sector. Yet, it should not be the university but the government that does the licensing negotiations because it would otherwise just increase the costs of patent lawyers at universities and distract researchers from what they do best, namely research and development.

This would not prevent researchers from becoming entrepreneurs themselves. In this case the technology patent would not be licensed out but used as the first asset of a new spin-off firm. The researcher would then entirely assume the role of an entrepreneur, who must ensure that the product can be sold on the market at a profitable price.

Often the private sector is discouraged from using a new prototype because inexperienced researchers overestimate the value of their invention and underestimate the fixed costs and risks that companies face when commercializing a new technology with uncertain market potential. The government that initiated the research initiative to achieve certain social and environmental objectives may have a real interest in encouraging the private sector to use the prototype and convert it into new goods and new markets. The government might therefore be willing to forego the licensing fee entirely in return for certain reservations when it comes to the commercialization of the product. It could ensure that there is ongoing research collaboration between the firm and the university and that the new products are affordable to the poor in developing countries (which may still be profitable for the company if it is free to charge a higher price in developed countries and able to prevent parallel imports).

If the prototype is still not attractive to the private sector because the market is too small to make a profit, the government can design additional incentives such as fast track regulatory approval and tax credits for product research. Once the private sector is willing to embrace the product because it expects to make a profit thanks to the additional incentives, it will be much more efficient and more end-user responsive than the public sector could possibly be.

5.4
Global Governance to Produce Global Public Goods

Often governments in developing countries may not have the means to offer sufficient incentives on their own to induce companies or research institutes to come up with products that would produce high social welfare triangles for their country. For example, improved orphan crops could save thousands of lives and significantly improve the health of the poor, but neither the local private sector nor the national governments in developing countries have the means and the know-how to successfully invest in such improvements. At the same time, multinational companies might have the know-how but do not have an incentive to invest (due the small market size for the good and the low purchasing power).

International donors could address these constraints by creating incentives for the private sector to produce such goods by offering a generous prize for the first company or research organization that was able to produce such a good [32] or contract an advance purchase that would boost expected demand [33].

Some people would denounce this as creeping privatization but the fact is that the new technologies that were derived from the information and biotechnology revolutions make it increasingly cost-effective to include the private sector in the management of public goods. Generally, these technologies permit smaller producers and more scope for competition [30].

5.5
Future Role of CGIARs in Promoting Agricultural Biotechnology in Developing Countries

From 1996 to 2004, biotechnology crops have reduced the volume of pesticide spraying globally by 6%, equivalent to a decrease of 172 500 t. The technology has also significantly reduced the release of greenhouse gas emissions from agriculture; a reduction equivalent to removing 5 million cars from the roads [34]. The increase in farm income that resulted from the adoption of GM crops is equivalent to adding 3–4% to the value of global production of the four main biotechnology crops [34]. Moreover the adoption of Bt cotton in many different developing countries turned out to have significant economic, health, and environmental benefits for small- and large-scale farmers alike [35]. All these facts just refer to GM crops and do not take into account the large economic and environmental gains that have been achieved by using all the other tools of modern biotechnology such as tissue culture, marker-assisted breeding, gene silencing, and genome mapping.

But why then do politicians often argue that agricultural biotechnology does not offer any benefits to society and the environment? This may be largely based on a generally hostile public opinion and vested interests that prefer the status quo in agriculture. Yet, it also seems that agricultural economists are not really able to provide convincing arguments about why agricultural biotechnology will also benefit the poor and the environment. Agricultural economists, who still stick to the principles of neoclassical welfare economics, assume that the private sector has, once again, no interest in addressing the needs of the poor and that therefore the public sector needs to step in and initiate a new public-sector-driven Green Revolution [36, 37]. The skepticism about private-sector involvement may be related to the general distrust of monopolistic competition that drives the process of technological innovation. In agriculture, it adds to the already existing scepticism related to the agricultural treadmill hypothesis, which treats technology as exogenous and implies that benefits from introducing technology in agriculture would not go to farmers but primarily to the seed and agrochemical industry and to the food consumers. This clearly contradicts the numbers of Brookes and Barfoot [34], who calculated an increase in global farm income through the adoption of GM crops of a cumulative total of $27 billion for the period 1996–2004, derived from a combination of enhanced productivity and efficiency gains. Obviously agricultural biotechnology must be more than just an agricultural treadmill. Moreover, it is wrong to reduce farmers to passive price-takers who struggle to survive in perfect competition. Farmers were always innovators and interested in collaborating with researchers; but national agricultural policies can either encourage or discourage innovative farmer activities.

5.6 Farmers as Innovators

The land grant college system in the USA was set up in the nineteenth century with the purpose of promoting applied science and stimulating economic activities in the rural areas. The state universities, which were established all over the country, had the explicit mandate to cooperate with local farmers and support their efforts in finding solutions to specific crop problems, but also to develop agricultural innovations with a commercial potential. This collaboration produced technological innovations, new agricultural products, and new companies in agribusiness. Apart from stimulating economic growth it also contributed to the social empowerment of rural areas in the USA. A similar development happened in Switzerland at the end of the nineteenth century. The first agricultural law was passed in 1893 with specific emphasis on improvement agricultural research and development and, in 1898, national agricultural research institutes were set up to meet this challenge [15].

This successful partnership between the university researcher and the farmer has largely been abandoned in Europe and the USA but New Zealand started to rediscover this success story after it decided to liberalize agriculture in the 1980s. The Royal Institutes of New Zealand were semi-privatized and its agricultural research activities only get funding if committed to making the farming sector more competitive and sustainable through innovation. This implies a close collaboration with business and the farming community. Even though genetically engineered crops are not yet approved for commercialization, agricultural biotechnology is at the center of this new agricultural policy in New Zealand. New Zealand's biotechnology industry generated an estimated revenue of NZ$811 million in 2005, with over NZ$250 million in exports (for further information see [38]). The industry has helped ensure the continued international competitiveness and efficiency of New Zealand's food and beverage sector. This focus on technological innovation did not just create a more sustainable and competitive agricultural sector (compared to the previous subsidy-based agricultural system) but also boosted the farmers self-confidence. Farmers do not feel victims of a new knowledge-based economy but an integral part of it [39]. The agricultural treadmill would have predicted a different outcome.

5.7 Crop Research Networks as a New Form of International Agricultural Research

Some would argue that New Zealand is an exception. It has invested a lot in knowledge and human capital, is well-governed, has excellent infrastructure and highly developed input and financial markets. Poor developing countries where none of this applies would face a much bigger challenge to make technology compatible with sustainable development, especially when it comes to the improvement of orphan crops that are largely grown by subsistence farmers. These farmers would first of all not benefit from private-sector innovations because there is no incentive to invest and, if they did invest, farmers would lack the knowledge to use new technology in a sustainable way.

The arguments may sound reasonable but they underestimate the power of creative solutions.

The Cassava Biotechnology Network (CBN), which started in 1988 as a global initiative to mobilize the development and application of biotechnology tools for the improvement of cassava agriculture, is an excellent example to illustrate how creative thinking can employ agricultural biotechnology for the benefit and empowerment of local subsistence farmers. Cassava is a typical orphan crop that is produced mostly by smallholders on marginal and submarginal lands in the humid and subhumid tropics. It is a starchy root crop that grows in a wide range of environments and is very tolerant to drought and acidic soils. Cassava has the advantage of flexible harvesting (the root tuber can be preserved in the soil for up to one year) and this makes it the crop of last resort for many poor farmers that prefer to harvest cassava whenever there is a shortage of food or animal feed.

CBN is based at the Centro International de Agricultura Tropical (CIAT) in Colombia and consists of a loose network of stakeholders that represent cassava research, cassava farming, cassava business, and international donors with an interest in cassava agriculture. The triennial CBN meetings serve as the major platform of information exchange. The aims are to share knowledge on cassava, identify new challenges in research, improve farmer adoption and marketing of cassava, and set up new research collaborations that are focused not primarily on research but on the development of new and useful products for cassava farmers. CIAT, which belongs to CGIAR research centers that were set up during the Cold War, is the major driving force behind this network. It responded to the financial crisis of the CGIAR system in the 1990s by basically reinventing itself as an engine of innovation in the area of orphan crops. CIAT realized that the old supply-driven system of international agricultural research was not really addressing the needs of the end-users and treated them in a rather paternalist way. It was a purely public sector initiative, which may be able to sponsor excellent research but does not know how to make useful products and disseminate them efficiently.

One goal of CIAT was therefore to get more involvement of the private sector and farmers to find out more about the effective demand for certain innovations in cassava agriculture. Once they have identified the areas of research that would meet the biggest demand, they look for the best partners worldwide to collaborate on joint research projects. Thanks to the advances in modern information and communication technologies, international research collaborations have become much cheaper and also more effective. This also explains why crop research networks such as CBN are probably best placed to facilitate an efficient exchange of knowledge and experience in the area of cassava research, production, and marketing worldwide. The interesting thing is that it is not just the development community but also the private sector and advanced research institutes in developed countries that have an interest in participating in such networks and in learning more about the advances in cassava science, the opportunities to create new markets out of cassava innovations, and ways to stimulate consumer demand for this neglected crop. Unsurprisingly, the fundraising activities for certain joint projects are not limited to the mainstream official donor community but also include the private sector and governments in developing countries. In fact, two thirds of the members of the CBN are from developing countries. This indicates that South–South technology transfer initiatives may eventually become as important as North–South technology transfer [40].

5.8
Agricultural Biotechnology as a Tool of Empowerment: The Case of the Cassava Biotechnology Network (CBN)

CBN is primarily designed to improve cassava as a food crop. Cassava is efficient in carbohydrate production but has a very low protein content. This causes a major problem of malnutrition if there is a high dependence on cassava-based daily meals. Agricultural biotechnology proved able to enhance the protein content of cassava [41] and was also used to analyze the biochemical pathways of β-carotene-rich cassava cultivated by indigenous tribes in the Amazon [42].

Another problem of food security is the creeping genetic erosion of cassava, which results in ever decreasing yields. Cassava is a vegetatively propagated crop and the planting material must therefore be exchanged in the form of cassava stakes rather than seed. Stakes are often highly contaminated with viruses and affected by genetic erosion. These problems largely account for the very low yields that cassava subsistence farmers reach in the field. Tissue culture techniques, some of the earliest tools of agricultural biotechnology, proved to be an excellent solution to address this problem. They allow the cheap and effective reproduction of clean cassava stakes. Moreover, tissue culture technology has been constantly improved over the past decades and the price of a tissue culture laboratory has dropped significantly over the past years. CIAT's Biotechnology Research Unit (BRU) has developed low-cost cassava in-vitro rapid multiplication techniques in collaboration with a Colombian farmer organization called FIDAR (Fundcación para la Investigación y Desarollo Agrícola). This comprises small tissue culture laboratories, cold chambers, and greenhouses, built mostly with local material. The use of local material made the end product six times cheaper than the official market version. Subsequently, men and women were trained to learn how to use their traditional knowledge about the best local cassava varieties and reproduce them in a tissue culture laboratory. The project proved to be very successful and induced especially women to set up local businesses and specialize in the local selling of high quality cassava stakes. In interviews conducted in 2003 with these women it was striking how this project also boosted their self-confidence. Suddenly, high technology ceased to be a magic practice that could only be handled by Western scientists but became a practical tool in daily life. It proves how the value of indigenous knowledge can be enhanced through the application of the new tools of agricultural biotechnology; and it shows that agricultural biotechnology can be a tool of empowerment [40].

CBN is, however, not just promoting cassava as a food crop but also as a cash crop. This is because cassava as a food crop hardly generates new markets as long as it only serves the immediate needs of those who produce it. Moreover, any production surplus that exceeds the demand of the family of the subsistence farmer and his neighbors is likely to turn into waste because of the absence of markets. The absence of markets is largely related to infrastructure problems and the lack of entrepreneurs who are interested in developing a market for cassava. Agricultural biotechnology has a potential to stimulate the development of new cassava markets by designing cassava varieties that are more productive and taste better. Moreover, the new techniques of agricultural biotechnology also have the potential to accelerate the bulking process of cassava, prevent its quick postharvest deterioration, and shorten the time of effective fermentation (detoxification). This would definitely help to overcome some of the constraints on cassava trade and attract more local entrepreneurs [40].

Most of the ongoing CBN research projects designed to make cassava a more attractive cash crop and increase consumer demand through better marketing are in collaboration with partners in the private sector. CBN is not interested in giving its product innovation away for free but wants the farmer to pay a price he can afford so that he learns to appreciate the value of innovation. The farmer's willingness to buy also signals that there is an actual demand for the product. Moreover, it changes the farmer's attitude: he is now taken seriously as a businessman and is not just a mere receiver of charity products.

All in all, the CBN projects prove that CGIARs can very well play a new role as brokers of new public–private partnership projects for the benefit of those who are most vulnerable in the knowledge economy. CBN could be still more successful if the Europeans could finally overcome their mental barrier towards agricultural biotechnology and support this success story.

6
Final Remarks

Current economic analysis of the value of agricultural biotechnology is still weighing the risks and benefits of existing GM crops for farmers and consumers. Even though, as shown in this article, these economic analyses largely confirm the economic and environmental benefits of existing GM crops, they largely ignore the fact that agricultural biotechnology advances at an unprecedented speed, continuously improving the economic and environmental performance of existing agro-biotechnology products and substituting earlier agrochemical products with new and cleaner biotechnology-based products. The value of this constant introduction of new goods into the economy is ignored by comparative–static welfare economics because the associated general and partial equilibrium models assume that all goods that could possibly exist do already exist. The fatal consequence is that the social welfare produced through the introduction of new goods is simply ignored. Welfare economics and agricultural economics in particular are therefore too much focused on the potential risks of new technologies and underestimate the benefits for society at large. The assumption of a comparative-static economy is especially flawed in the new knowledge economy where an exponential growth of knowledge leads to an exponential increase of the probability that new goods come into being. This largely explains why national agricultural policies that still rely on the principles of classical welfare economics have largely discouraged farmers from becoming more competitive and innovative (e.g., in Europe) while new policies that have embraced the ideas of new growth theory are encouraging farmers to become innovative and competitive (e.g., in New Zealand). New growth theory policies also proved to be more effective in improving the environment and social welfare.

If new technologies are not only to serve the attractive markets in affluent societies but also to contribute to a better life in poor developing countries, it is time to learn from past mistakes and also to design new development policies that take into account bottom–up solutions. As shown in the case of the Cassava Biotechnology Network (CBN), agricultural biotechnology proved to be perfectly compatible with such bottom–up solutions that involve the local farmers as well as the local private sector. Moreover, modern information and communication technologies offer new forms of decentralized international collaboration that enable a stronger involvement of local participants and a more effective international network of research and product development. In this context, CBN represents another example of a new approach that is no longer based on the old principles of welfare economics but has embraced new growth theory and thus enabled the marginal farmers in developing countries to participate more effectively in the new knowledge economy.

References

1. Lovejoy AO (1936) The great chain of being: a study of the history of an idea. Harvard University Press, Cambridge
2. Romer P (1990) Endogenous technological change. J Pol Econ 98(5):71–102
3. The Economist (2006) The battle for brainpower: a survey on talent. October 7th, 2006
4. Harberger A (1959) Monopoly and resource allocation. Am Econ Rev 49:134–146
5. Schumpeter JA (1942) Capitalism, socialism and democracy. Harper & Row, New York
6. Samuleson P (1947) Foundations of economic analysis. Harvard University Press, Cambridge
7. Coase R (1974) The lighthouse in economics. J Law Econ 17:357–376
8. Arrow K (1951) Social choice and individual values. Yale University Press, New Haven
9. Buchanan JM, Tullock G (1962) The calculus of consent: the logical foundations of constitutional democracy. University of Michigan Press, Ann Arbor
10. Solow RM (1956) A contribution to the theory of economic growth. Quart J Econ 70:65–94
11. Romer P (1994) New goods, old theory, and the welfare costs of trade restrictions. J Dev Econ 43:5–38

12. Aerni P (2006) The welfare costs of not being part of the knowledge economy: why rural development needs more creative policy strategies. ATDF J 3(4):27–47
13. Warsh A (2006) Knowledge and the wealth of nations. Norton, New York
14. Tullock G (2005) The organisation of inquiry. Liberty Fund, Indianapolis
15. Rieder P, Anwander Phan-Huy S (1994) Grundlagen der Agrarpolitik. Hochschulverlag, ETH Zürich, Zürich
16. Henrichsmeyer W, Witzke HP (1994) Agrarpolitik 1/2. Eugen Ulmer, Stuttgart
17. Cochrane WW (1979) The development of American agriculture: a historical analysis. University of Minnesota Press, Minneapolis
18. The Economist (2005) The poorest part of America. December 8th, 2005
19. Whitfield J (2006) Agriculture and environment: how green was my subsidy? Nature 439:908–909
20. Anderson RS, Levy E, Morisson BM (1991) Rice science and development politics. research strategies and IRRI's technologies confront Asian diversity (1950–1980). Clarendon, Oxford
21. Desta MG (2002) The law of international trade in agricultural products: from GATT 1947 to the WTO Agreement on Agriculture. Kluwer, London
22. Aerni P (1999) Public acceptance of transgenic rice and its potential impact on future rice markets in Southeast Asian countries. Dissertation 13471, ETH Zürich, http://www.iaw.agrl.ethz.ch/people/Personen/aernip/Publications/Enfassung.pdf (last visited: 28 Feb 2007)
23. Byerlee D, Morris M (1993) Research for marginal environments: Are we underinvested? Food Pol 18(5):381–393
24. Pollack A (2001) Changing times challenge the World Hunger Organization. The New York Times, Tuesday, 15 May, Science Section
25. Gupta A (1997) Honey Bee Network: Linking knowledge-rich grassroots innovations. Development 40(4):36–40 http://www.sristi.org/cms/publications, (last visited: 28 Feb 2007)
26. United Nations Development Program (UNDP) (2001) Human development report: Making new technologies work for human development. Oxford University Press, New York
27. Mowry DC, Nelson RR (1999) Sources of industrial leadership: studies of seven industries. Cambridge University Press, New York
28. Margolis M (2005) Flying south. Newsweek Magazine, December 28, pp 44–46
29. Center for Strategic and International Studies (CSIS) (2006) Biotechnology and agriculture in 2020. CSIS, Washington DC
30. Heal G (1999) New strategies for the provision of public goods. In: Kaul I, Grunberg I, Stern MA (eds) Global public goods: international cooperation in the 21st century. Oxford University Press, New York
31. Rosenberg N (2000) Schumpeter and the endogeneity of technology. Routledge, NY
32. Masters W (2005) Research prizes: a new kind of incentive for innovation in African agriculture. Int J Biotechnol 7(1/2/3):195–211
33. Glennerster R, Kremer M, Williams H (2006) Creating markets for vaccines. Innovations 1(1):67–79
34. Brookes G, Barfoot P (2005) GM crops: the global economic and environmental impact – the first nine years 1996–2004. Agbioforum 8(2/3):187–196
35. Aerni P (ed) (2005) Agricultural biotechnology in developing countries: perception, politics and policies. Int J Biotechnol 7(1/2/3):1–6
36. Conway G (1999) Doubly green revolution: food for all in the 21st century. Cornell University Press, Ithaka

37. Cooper J, Lipper LM, Zilberman D (eds) (2005) Agricultural biodiversity and biotechnology in economic development. Springer, Berlin Heidelberg New York
38. New Zealand Trade and Enterprise (2006) Biotechnology and agritech - growth industries. NZTE http://www.nzte.govt.nz/section/13202.aspx (last visited: 28 Feb 2007)
39. Aerni P (2006) Agricultural policy in New Zealand: Making global competitiveness compatible with environmental sustainability. CIS News, ETH Zurich, pp 7–9
40. Aerni P (2006b) Mobilizing science and technology for development: the case of the Cassava Biotechnology Network. Agbioforum 9(1):1–14
41. Peng Z, Jaynes JM, Potrykus I, Gruissem W, Puonti-Kaerlas J (2003) Transgen Res 12(2):243–250
42. Carvalho LCB, Thro AM, Vilarinhos AD (eds) (2000) Cassava biotechnology: proceedings of IVth International Scientific Meeting of the Cassava Biotechnology Network (CBN-IV), 2–6 Nov 1998, Salvador, Brazil. Embrapa, Brasilia

Invited by: Professor Sautter

Adv Biochem Engin/Biotechnol (2007) 107: 97–112
DOI 10.1007/10_2007_046

Published online: 31 March 2007

Exploration and Swiss Field-Testing of a Viral Gene for Specific Quantitative Resistance Against Smuts and Bunts in Wheat

Thomas Schlaich · Bartosz Urbaniak · Marie-Laure Plissonnier · Nicole Malgras · Christof Sautter (✉)

Institute of Plant Science, Swiss Federal Institute of Technology Zurich, Universitätsstr. 2, 8092 Zurich, Switzerland
christof.sautter@ipw.biol.ethz.ch

Abstract The viral gene for the killer protein 4 (KP4) has been explored for its antifungal effect in genetically modified wheat to defeat specifically the seed-transmitted smut and bunt diseases. In vitro both important seed-transmitted diseases of wheat, loose smut (*Ustilago tritici*) and stinking smut (*Tilletia caries*), are susceptible to KP4, whereas all other organisms tested so far proved to be not susceptible to KP4. For studies in planta we used stinking smut as a model fungus. In greenhouse experiments, two KP4-transgenic wheat lines showed up to 30% lower symptom development as compared to the non-transgenic control. As the last step in the proof of concept, field-testing has shown for the first time increased fungal resistance of a transgene in wheat. Due to its specificity against smuts and bunts, KP4 presents a very low risk to humans and the environment. Field-testing in Switzerland is regulated by a strong law, which for research is acceptable if legally and scientifically correctly applied.

Keywords Field-testing · Fungal defense · Genetic modification · Killer protein · Wheat · Smuts and bunts

1 The Scientific Approach

1.1 Introduction

Besides rice, wheat (*Triticum aestivum*) is the most important staple crop plant. Wheat represents an attractive target for many fungal pathogens [1], and fungal diseases cause devastating losses in wheat worldwide [2, 3]. Fungal diseases can be controlled by agronomic practice, including chemical spray, or by genetic improvement. Breeding efforts focused for a long time on resistance genes providing qualitative resistance according to the "gene for gene" concept of Flor [4]. Quantitative resistance traits are more difficult to breed, since quantitative resistance genes have to be pyramidized and their antifungal properties often depend on synergistic effects with other resistance genes. This makes it difficult following the phenotype of these genes. Thus, during the breeding process molecular methods and sophisticated statistics are required [5].

Not all pathogens and their respective strains can be controlled by means of the resistance genes in the wheat gene pool [6]. Potent chemical fungicides are available but are not always desirable ecologically in view of their side effects, e.g., impairment of mycorrhiza [7], effects on soil microbes [8], or reduction of predator efficiency [9, 10]. An alternative or supplementary approach is genetic transformation. This can confine the chemical control to the tissue of the host plant itself, and thus represents an ecological advantage over spraying into the environment. However, most projects of the gene transfer approach used broad spectrum fungicidal proteins like chitinases, glucanases, or ribosomal inhibiting proteins [11–16]. In approaches with broad spectrum antifungal transgenes, side effects are reduced, but useful fungi like mycorrhiza still might be affected as well.

We explore a system of specific, quantitative resistance in genetically modified wheat, which should affect only pathogenic fungi. The so-called interstrain inhibition system of *Ustilago maydis* offers several viral genes with antifungal properties. These viral genes are nonhomologous, and act specifically only against members of the order *Ustilaginales*, which contains exclusively plant pathogenic fungi. This approach offered the chance to study the promise of such specific, quantitative resistance. This system has the potential of fungal defense with a narrow action spectrum defeating exclusively pathogens and at the same time having possibly no side effects on putative human consumers or on the environment.

1.2 The Interstrain Inhibition System

Corn smut, *Ustilago maydis* (de Candolle) Corda, exists in a multitude of different strains. Some 5% of the naturally occurring collections contain viruses, which are cytoplasmically inherited and are the causative agent of interstrain inhibition in *U. maydis* [17]. This interstrain inhibition has been called the "killer" phenomenon [18] in analogy to an earlier discovered system in yeast [19]. At least three different killer specificities are known in *U. maydis* [20] of nonhomologous modes of action [18]. The viruses contain a three-partite genome of double-stranded RNA, from which two parts share homology [21]. One part encodes for a protein gene, which is transcribed and translated by the fungal host. The protein is secreted into the apoplast of the plant host, where it inhibits growth of fungal hyphae of the order *Ustilaginales*, which do not contain this virus [22] (for an explanation, see Fig. 1). It has been reported that killer protein 4 (KP4) binds specifically to and inhibits the regulation of an L-type Ca channel in the plasma membrane of the fungus [23]. This does not kill the cells but inhibits longitudinal growth of the hyphae. Addition of external Ca^{2+} or cAMP can compensate for the KP4 effect and *U. maydis* recovers [23]. The same effect, but quantitatively less pronounced, was shown with *Tilletia caries* (Plissonnier ML (2003) Diplomarbeit ETH Zurich/Univ. de Bourgogne, unpublished results). The cAMP pathway is involved in the growing modus of *U. maydis* [24].

Naturally in this system, the hyperparasite (i.e., the virus) protects its host (i.e., *U. maydis*) from competition by other fungal strains of related species in the same host plant [22]. The virus-infected *U. maydis* strain is protected

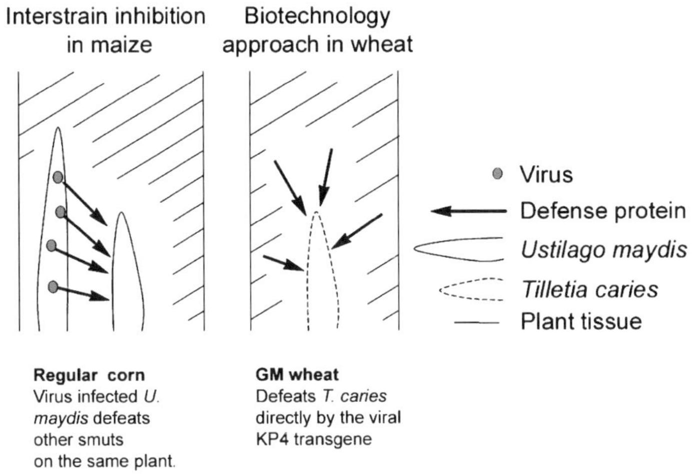

Fig. 1 Interstrain inhibition in maize and biotechnology approach in wheat

against its "own" toxin by a recessive mutation in the receptor protein complex (for details see [25]). This mutation might cause a drawback for the vigor of the fungus and explain the low selective advantage for the virus-infected *U. maydis* strains, illustrated by only 5% of the *U. maydis* population containing these viruses.

Two of the three known [19], nonhomologous genes from three different viruses (KP4 and KP6) are available in our laboratory as a generous gift from Dr. J. Bruenn (Buffalo, USA). We explored this interstrain inhibition system from *U. maydis* viruses for increasing resistance against smut fungi in wheat. For that purpose we transferred the KP4 gene into the two Swiss spring wheat varieties Golin and Greina (Fig. 1), which are particularly susceptible to stinking smut [26].

1.3
Seed-Transmitted Diseases

Pathogens that produce their diaspores (i.e., distribution units, such as fungal spores) in seeds of crop plants cause a particular phyto-sanitary problem: seed material harvested from infected fields will spread the disease during sowing. A number of fungal pathogens belong to these so-called seed-transmitted diseases. Among these fungal diseases are the five smut and bunt species: loose smut (*U. tritici*), stinking smut (*T. caries*), common bunt (*T. laevis*), dwarf bunt (*T. controversa*), and Karnal bunt (*Neovossia indica*), of which *T. laevis*, *T. caries*, and *T. controversa* are closely related. Flag smut (*Urocystis agropyri*) has also been described as a seed-transmitted fungal disease [27]. If these diseases are not controlled they will accumulate quickly, particularly when a farmer repeatedly uses part of his own harvest for sowing. Therefore, farmers often prefer to use certified seed material from seed producers rather than part of their own harvest, as they had to do in ancient times and many farmers in developing countries still have to do. For some 50 years these diseases have been controlled by chemicals which effectively prevent the outbreak of these diseases, but at the same time contribute to the undesired chemical input in agronomy.

Seed-transmitted diseases can also spread by shipment of harvests for food or feed use which cannot be treated by chemicals. This is a major threat to countries and regions from which they are at present absent [28]. Therefore, international trade agreements regulate seed-transmitted diseases not only for sowing material but also for shipment of harvests designated as food [29]. Severe outbreaks of a seed-transmitted disease usually lead to quarantine measures, i.e., the farmers are not allowed to sell their harvest outside of the epidemic territory. In addition, they have to sterilize their harvesting machines in order to prevent the spread of the disease [30].

Recent developments favor a lower chemical input for agriculture. This includes organic farming and, in general, modern agronomic practices under the paradigm of sustainable agriculture and land use. Organic farmers, for example, have control measures against stinking smut (Heinzer L (1998) Diplomarbeit, FAL Reckenholz, unpublished results) which have to stand the test over the years, but no control for loose smut [31]. For reduction of chemical input in disease control, genetically improved plant varieties are the method of choice, provided they are available. Due to the genetic flexibility of the pathogens, this is never sufficiently the case and requires continuous research in this field.

In this project, stinking smut served as a simple and safe model to test whether the genetically modified plants provide improved fungal resistance against smuts and bunts. In contrast to loose smut, stinking smut does not spread its spores before thrashing. The results with stinking smut might be applicable to other wheat-infecting smuts and bunts.

1.4
Biotechnology Approach

We constructed a chimeric gene for expression of KP4. We used cDNA of KP4 and inserted this cDNA into an expression vector under the control of the maize ubiquitin promoter including its intron and the 35S poly(A) signal. We transferred the complete pUC 19 vector containing this insert into the wheat plants. The selection marker was introduced by cobombardment with the *bar* gene under the rice actin promoter (for further details, see [32]).

The varieties Golin and Greina have been chosen according to their susceptibility against stinking smut, neglecting their low transformation efficiency. In fact, both these varieties proved to be fairly recalcitrant to transformation. We managed to regenerate two KP4 Golin lines and four KP4 Greina lines from a total of 559 precultured, immature embryos. From these six lines, which contained the gene according to Southern blot analysis of the genomic DNA, one Golin and one Greina line exhibited specific antifungal activity against the model fungus *U. maydis* in vitro. This corresponds to a transformation efficiency of 0.5%. This is low as compared to model varieties, which allow for transformation rates of up to 10% and more [33]. The two KP4 lines Golin 5 and Greina 16 inherited the transgenes in a Mendelian way. We continued our study with nonsegregating progeny lines from the primary regenerates Golin 5 and Greina 16 [32].

1.5
Results from In Vitro Analyses

Antifungal activity can be measured by a specific and quantitative growth inhibition test using diffusion zones on an agarose surface (Fig. 2). The antifungal effect is specific against those strains of *U. maydis* which do not contain the KP4-encoding virus. The *U. maydis* strain which contains the KP4

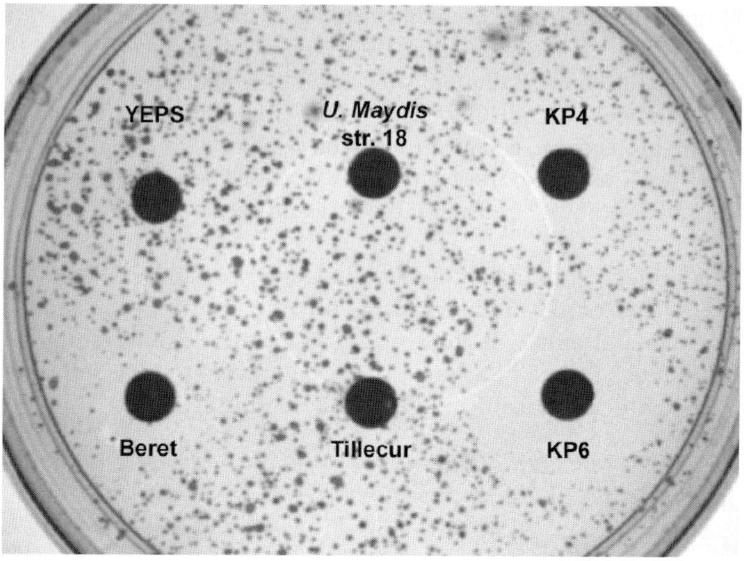

Fig. 2 Antifungal activity measurement by growth inhibition zones on agarose diffusion tests. *U. tritici* (loose smut, *gray dots*) on the surface of an agarose plate. Note the different sizes of the inhibition zones. *Black*: filter paper soaked with the respective compound; YEPS: medium of the fungal culture; *U. maydis* str. 18: *Ustilago maydis* strain 18, which does not contain a virus and thus is susceptible against KP4 and KP6. Beret: fungicide (active compound Fenpiclonil); Tillecur: yellow mustard powder (recommended for stinking smut control by ecological farming in Switzerland)

encoding virus and KP4-free segregant plants served as negative controls. This characteristic pattern of antifungal effects excluded nonspecific smut resistance in the genetic background and confirmed the specific action pattern of the KP4.

Since *U. maydis* is not a pathogen of wheat, it was important to test also wheat pathogens, such as stinking smut (*T. caries*) or loose smut (*U. tritici*). Both fungi can hardly be grown on media in the absence of the host plant. However, we optimized the growing conditions in vitro and managed to apply the agarose diffusion test as well to these two biotrophs.

We tested mixtures of strains from natural collections kindly provided by Dr. G. Schachermayr and Dr. H.-R. Forrer (Federal Research Station FAL, Reckenholz, Switzerland). For *T. caries* we had in addition strains of defined virulence available by courtesy of Dr. B. Goats (USDA, Aberdeen, ID, USA). In vitro, purified KP4 from the fungal supernatant [34] exposed antifungal activity against all isolates and wild-strain collections in the diffusion test.

Expression of the transgenes *kp4* and *bar* in the transgenic plants was measured as the physiological activities of the respective proteins, the antifungal activity of KP4, and phosphinotricin acetyl transferase (PAT). In planta, the KP4 accumulated in KP4 Greina and KP4 Golin with age, as measured by the diffusion test using homogenate from plant tissue. The antifungal activity is maximal in seedlings, where the fungi infect, and in developing ears, where the fungi sporulate. The expression of the *bar* gene can be detected by Basta spraying, since only transgenic plants are resistant to phosphinotricin, the active compound of Basta. The antifungal activity of the KP4 gene was inherited by cosegregation together with the *bar* gene up to the seventh generation.

1.6 Greenhouse Experiments

In order to test the antifungal effect against *Ustilaginales* in planta, we infected plants artificially by mixing fungal spores with kernels and planted these in pots in the greenhouse [32]. Table 1 shows that both KP4 lines exhibited a reduction in stinking smut disease symptoms of up to 30% compared to wild-type controls or null segregants [32, 35]. Since KP4 is expected to act like a quantitative resistance gene, we also tested different infection pressures, which we applied by different spore concentrations. Indeed, we found the best antifungal effect of the KP4 plants in the region of medium response values of the relation between infection pressure and symptom development [34].

Table 1 Improved fungal resistance in planta after artificial infection with *U. tritici*

	Average number of ears per plant	Percentage of diseased ears	Difference %
Greina 16	2.73	43	31
Greina wt	3.00	74	
Golin 5	3.46	67.5	32.5
Golin wt	3.75	100	
LSD 5	0.35		15

LSD 5: least significant difference at 5% error probability

2 Field-Testing in Switzerland

2.1 Field Test Application and Biosafety

As plant response to fungal infection depends upon environmental conditions, it is necessary to verify greenhouse data in the field. In this project we considered the field test as the last experiment for proof of concept. We applied for the field test first in October 1999, right after the greenhouse experiments. We performed a variety of tests for biosafety concerning the KP4, in order to allow for a risk assessment, which is required for the field test application. All these results were part of the application procedure and have been included in the public information on this experiment. These results are available in German on the website www.feldversuch.ethz.ch.

Toxicity was tested with a hamster cell line, an insect cell line, and a human kidney cell line for susceptibility against KP4 (Fig. 3). We were not able to detect any significant effect of the KP4 in all the tested cell cultures [34]. Rat neonatal heart cells are known to contain L-type Ca channels, which are the target of KP4 in the plasma membrane of the fungal hyphae. Incubation of these cells in culture with purified KP4 did not change detectably the shape of the cells, their sarcoplasmic reticulum, or their cytoskeleton [34].

Allergenic properties were studied with two approaches. Firstly, we searched for known allergenic amino acid sequences [36] and did not find any. Secondly, we performed a preliminary test for the persistence of KP4 in an artificial stomach fluid, which might provide an indication of putative allergenicity [37]. After some 8 min in the artificial stomach fluid all antifungal activity of the KP4 was lost according to the diffusion test, and the protein was no longer detectable on a Coomassie blue stained SDS-PAGE electropherogram [34]. These results indicated a very low chance of KP4 being allergenic.

Horizontal gene transfer from plants to microorganisms in the soil has been considered an extremely rare event, which has never been observed in nature [38] but has been an issue in the evaluation of the field test application. Fragments of the ampicillin resistance gene have been inserted together with the KP4 and the *bar* gene. The concern was that these fragments could be inserted into the genome of a soil bacterium, and restore there a defective ampicillin resistance gene. We tested the number of naturally ampicillin-resistant bacteria present in soil. About 5% of the aerobe and mesophile soil bacteria, which can be cultivated in vitro, are naturally ampicillin resistant due to unknown mechanisms. We calculated some one billion naturally ampicillin-resistant bacteria per square meter of soil. This makes horizontal gene transfer an event of no account—if it ever happens.

Crossing out via pollen into neighboring cultures is of no significance in wheat, since wheat is a strong self-pollinator [39]. Out-crossing beyond a few meters has only been observed with male sterile acceptor plants up to a maximum of 48 m in a few rare cases [40–42]. However, pollen flies and even if the pollen does not fertilize, it might affect passersby. We did not want to introduce this little risk into the neighborhood, even though there was negligible chance of any toxic or allergenic effect. Therefore, we prevented pollen flow completely. We covered the plots during flowering of the plants with small, pollen sealing tents (Fig. 4). Theoretically, these tents should allow air and moisture to diffuse and light to pass. In fact, about 90% of the light intensity was ab-

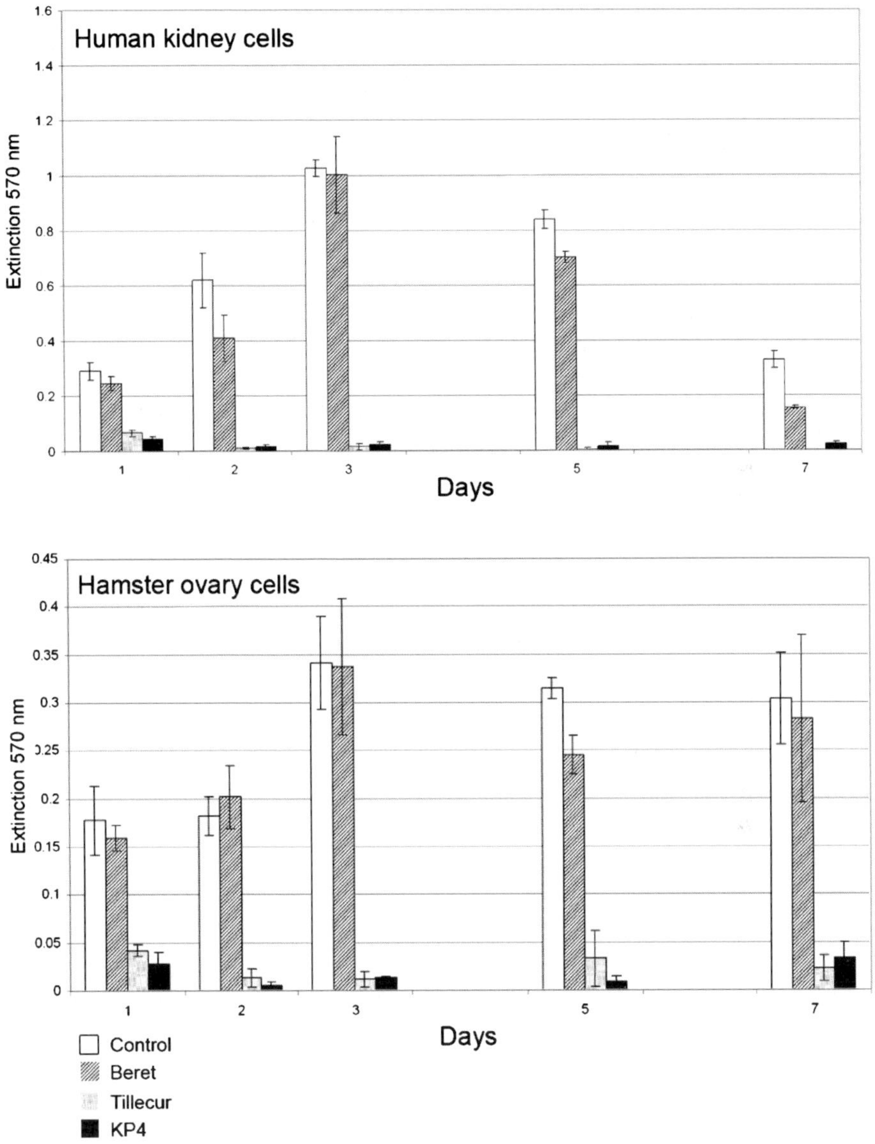

Fig. 3 Comparative viability of mammalian cells in the presence of KP4. *Top*: human kidney cell line (HEK 293) and *bottom* hamster (CHO-K1) cells have been tested for effects on viability in the presence of KP4 in terms of cell number and respiration. We found no significant difference in viability in KP4-treated cell cultures compared with the controls. From the effect of the nontoxic compounds Beret (conventional chemical seed control) and Tillecur (stinking smut control suggested by ecological farmers) in the viability test, we concluded that KP4 must also be nontoxic

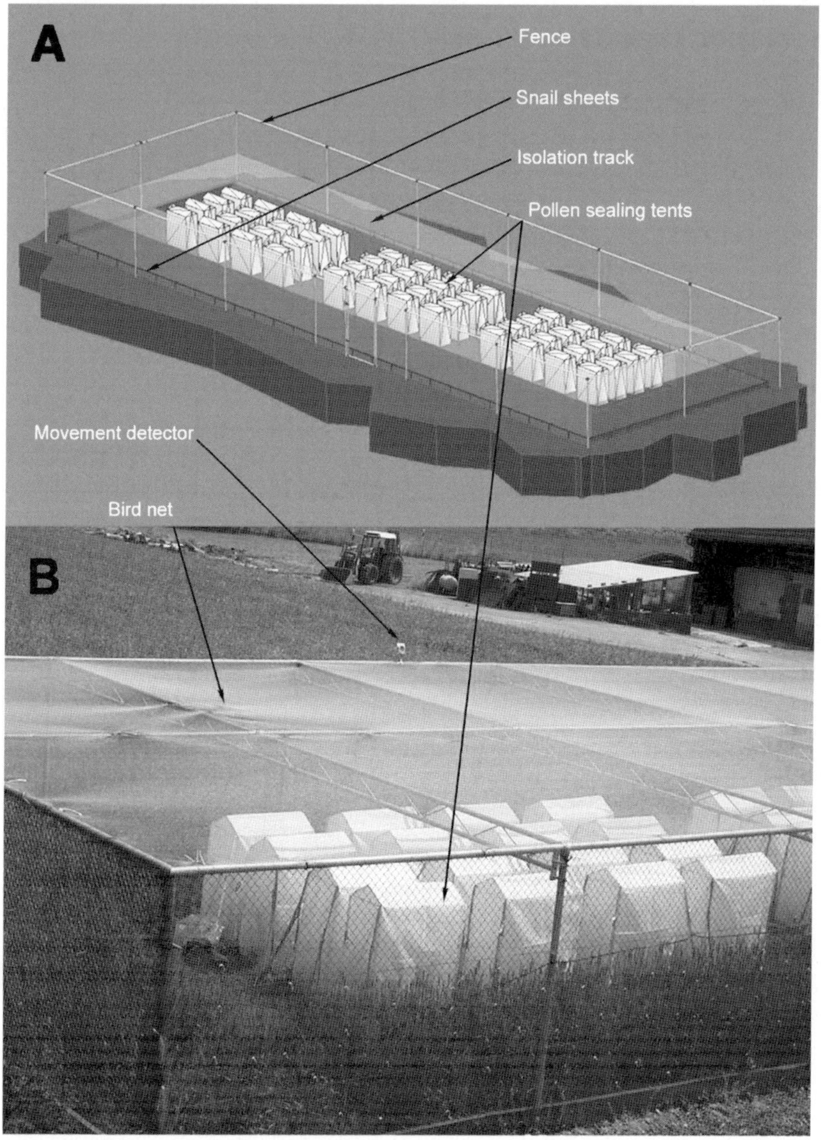

Fig. 4 Safety measures at the field test site

sorbed by the tissue and moisture was high due to low air exchange through the tissue. The atmosphere under these tents was completely calm. If plants are grown exclusively under these tents, it might lead to low lignification. However, in the case of our field test the impact of the tents on plant and disease development was low. We installed the tents for only 3 weeks shortly before pollination started, too late to influence the development of plant or disease.

Further safety measures were (Fig. 4): snail sheets against small rodents; a 2 m-high fence with a locked door, although the experimental site was inside the area of the field station, which was surrounded by a fence itself; a bird net; a 2 m-wide isolation track with nongenetically modified (wild-type) wheat, movement detectors, and a video camera, in order to observe the field day and night. Since we experienced pressure from anti-gene technology activists, we stayed at night close to the field site during the first 3 weeks after sowing, in order to prevent any theft or other removal of seeds. After the seeds were germinated, we hired a professional guard.

Table 2 Biosafety experiments

Experiment	Collaboration
Frequency of amp, KP4, and *bar* gene in excrements of Oulema sp.	Dr. Leo Meile (Inst Food Sci., ETHZ)
Frequency of amp, KP4, and *bar* gene in *Septoria nodorum*	Prof. B. McDonald (Inst. Plant Sci, ETHZ) and Dr. G. Vogel (KL, Basle)
Pleiotropic expression of endogenous genes upon fungus infection in genetically modified wheat	Own experiments
KP4 gene flow from wheat to *Aegilops cylindrica*	Prof. F. Felber (Univ. Neuchatel)
Field mycorrhiza infection frequency in KP4 wheat (anatomical study)	Dr. J. Jansa (Inst. Plant Sci., ETHZ)
Aphids on KP4 wheat in the field test	Dr. J. Romeis (FAL Reckenholz)
Oulema on KP4 wheat in the field test	Own experiments
Scoring for non-smut fungal diseases	Dr. M. Winzeler (FAL Reckenholz)
Putative impact of KP4 wheat on soil bacteria	Dr. F. Widmer (FAL Reckenholz)
Natural background of ampicillin-resistant bacteria in the soil	Dr. Leo Meile (Inst. Food Sci., ETHZ)
Out-crossing of wheat into male sterile wheat in the isolation track	Own experiments
Persistence of KP4 DNA in the soil	Dr. G. Vogel (KL, Basle)
Aphids on KP4 wheat in the greenhouse	Dr. G. Lövei (Flakkebjerg, Denmark)
Mycorrhiza in the greenhouse in KP4 wheat, physiological study	Prof. A. Wiemken
Effect of pollen sealing tents on out-crossing	Prof. F. Felber (Univ. Neuchatel) Dr. G. Vogel (KL, Basle)
Worst-case scenario for an impact of a KP4 horizontal gene transfer event	Dr. Leo Meile (Inst. Food Sci., ETHZ)
Effect of heat treatment of the soil on DNA persistence	Dr. G. Vogel (KL, Basle)

Swiss law requires additional biosafety experiments in parallel and connected to the field test. These biosafety tests have to contribute to knowledge about the biosafety of genetically modified plants, but do not necessarily need to concern the particular transgenic event under study. We arranged a number of accompanying biosafety experiments (Table 2). Most of these experiments were performed either by our own group or in collaboration with a number of colleagues. All but one of these experiments have been evaluated and confirmed the very low risk of the field test and thus our risk assessment. One experiment could not be finished, since we could not inspire an undergraduate student for the topic to perform the project for a diploma thesis. We could not hire additional personnel due to lack of financial support, as these additional experiments were not funded.

2.2 Legal Efforts and Timetable

The field test was performed long after the first greenhouse results were published [32]. Actually, we (CS) applied for permission for the field test for the first time in October 1999 with the title "Microbial interaction of KP4-GMO Swiss wheat varieties." This first application was not reviewed by the authorities, since the principal investigator (CS) was not accepted by the Swiss Federal Office for Environment, Forest, and Landscape (BUWAL) as a member of a federal research and teaching institution, but was treated as a private person. Private persons are required to deposit 20 million Swiss francs (ca. US $15 million) for a field test application to provide liability fees for putative damage to health or environment.

The Institute of Plant Science was ready to apply for the field test, but asked for revision of the application. We removed the biosafety experiments and changed the title to "Field performance of genetically modified KP4 wheat varieties". The revised version of the application was handed in to the BUWAL in November 2000, and after some negotiations was finally accepted for reviewing by January 2001. In February 2001 we informed the public neighborhood of the test site. In March 2001 the authorities made 27 additional requests, many of them not relevant for a small-scale test plot of only 8 m^2 and 1600 genetically modified plants, but appropriate for commercialization. We answered these questions and reintroduced and complemented the biosafety experiments into the application.

Finally, in November 2001 the application was refused. The main reasons given to justify the refusal were: (1) the plants contained fragments of the ampicillin resistance gene; (2) molecular data were not sufficient; (3) pollen sealing tents are not storm resistant; and (4) the effect of the transgene on the environment has not been tested. It has to be pointed out that all the committees, offices, and external reviewers suggested the experiment be permitted under certain conditions. Actually, five members of the Swiss biosafety committee resigned upon this decision of the BUWAL [43].

Considering the justification for the refusal, we saw a good chance for an appeal, since in a small-scale test plot: (1) the ampicillin fragments pose no particular risk for biosafety; (2) the desired additional molecular data are not required for risk assessment; (3) although the tents were not required to prevent out-crossing, we had tested the tents beforehand in a wind channel and had shown that they are indeed storm proof up to 120 km/h wind velocity, corresponding to the wind velocity at 20 m height during the most devastating storm of the last few decades in the area; and (4) the effects of the transgene on the environment can be tested only in the environment, and therefore can hardly be a general argument against a first small-scale field test in the environment.

In September 2002 the appeal was granted in all points by the executive department of the Swiss government (Department for Environment, Energy, Traffic, and Communication, UVEK). The case was sent back to the BUWAL for a new decision.

In the middle of December 2002 we reminded the BUWAL of the pending issue. The BUWAL permitted the experiment end of December 2002 under certain conditions. This permission was appealed by nine inhabitants of the village in the neighborhood of the test site. These opponents were supported by Greenpeace which, together with the Swiss ecological farmers union, also appealed up to the Swiss Supreme Court. In March 2003 the Swiss Supreme Court found a legal gap in the Swiss ordinance on release of genetically modified organisms, developed by the BUWAL. The BUWAL had to change this release ordinance before we could apply again in 2003. The experiment was then permitted in October 2003 and immediately challenged by an appeal of the same opponents (see above). We finally obtained valid permission for the field experiment in spring 2004.

The experiment was started on 17 March 2004 and finished according to schedule in July 2004. A visit by Greenpeace on 26 March posed a severe threat to biosafety, but fortunately did not harm the safety measures, in contrast to the year before. In 2003 Greenpeace activists had destroyed the safety installations of the empty test site and threatened employees on the research station. A few Greenpeace activists were always present at the test site throughout the duration of the field test. We never knew whether Greenpeace might again act violently against the experiment. Therefore, we hired a professional guard to observe the place every night and at the weekends.

The data of the experiment are evaluated and the summarized results are published [34]. In brief, the field test confirmed the results from the greenhouse tests. Both KP4-transgenic wheat varieties showed a 10% lower rate of symptom development as compared to the wild-type controls. For a quantitative resistance gene 10% is considerable, since natural quantitative resistance genes often contribute by less than 10% to pathogen defeat [5]. Our publication is the first reported field resistance against a fungal disease of genetically modified wheat. None of the results of biosafety tests justified any of the concerns brought up during the evaluation of the application.

It should be pointed out that we feel serious environmentalists perform a necessary role in society and we also appreciate the need for a strict law.

2.3 Expenses

The high level of public awareness, which turned the experiment into such a political issue, had in particular two consequences hampering research in this field: one was a financial problem and the other a handicap in competitiveness.

2.3.1 Financial

The project was funded by the Swiss National Foundation (SNF) till end of 2000. The field test performed in 2000 would have perfectly fitted into this schedule. After that, the project was kindly prolonged by the SNF until the end of 2003, when this biotechnology program, finally, ran out of financial resources. During its last year (2004) the project was financially supported by the Swiss Federal Institute of Technology Zurich. The financial expenses included not only the collaborator in this project, who was absolutely necessary to maintain the expertise, but also the additional safety measures and their repair after destruction by Greenpeace. In addition, we incurred expenses for lawyers. Without their help our appeal would have had little chance. The professional guards during the night and the weekends had to be paid and we had to reimburse the Cantonal Laboratory in Basle (KLB) for the PCR tests to detect the KP4 gene fragments in the soil. PCR from soil material is not trivial, since the humin acids in the soil inhibit the polymerase in the PCR. The KLB is particularly experienced in detection of DNA by PCR from soil. It was somewhat incongruous that we had to pay the legal costs of the Supreme Court in 2003, since we formally lost the case, although the sentence was due to a failure in the release ordinance, which was beyond our control. In total more than 600 000 Swiss francs (corresponding to about US $500 000) had to be spent in addition to the original regular 3-year grant for one postdoctoral fellow. This calculation does not include the expenses of the police, who had to protect the test site several times against protesters.

2.3.2 Competitive Expenses

The second expense was caused by the time delay and the public awareness. The time delay makes it impossible to keep a postdoctoral fellow interested in the project. A postdoctoral researcher has to publish, and cannot afford to wait 4 years for a valid decision of the authorities about a single experiment. Instead of research, a lot of time had to be spent on writing appeals, answers to replies to appeals, and other legal papers up to a total of more than 500 pages. Nevertheless, little of the scientific content of these papers is useful for research publications, which contribute to the scientific reputation, but had to be published anyway either via the BUWAL or on the website, in order to meet the public demand for transparency. Although the website is in German, it is a disclosure also to competitors in the field. In other countries the legal process is more straightforward. An application for the same experiment with the same plants took three pages and 6 weeks to be permitted in the USA. A well-organized laboratory could have done our complete project from scratch in the USA within the 4 years delay caused by the legal process in Switzerland.

3 Conclusion

If the current situation continues, plant sciences in Switzerland will have to deal with a de facto moratorium on field tests in basic research. This unofficial ban also includes biosafety research. Probably, Swiss plant research will survive this situation, but projects with any kind of application perspective—for which at least a small-scale field test is required as part of the proof of concept—will become very difficult.

Acknowledgements We greatly acknowledge legal advice from Dr. Stefan Kohler (Pestalozzi Lachenal Patry, lawyers, Zurich). The project was funded by the Swiss National Foundation and the Swiss Federal Institute of Technology Zurich (ETHZ). We thank Prof. W. Gruissem (Institute of Plant Biotechnology, ETHZ), Dr. S. Bieri (ETH-Rat), and Vice Presidents Prof. A. Waldvogel and Prof. U. Suter for their support.

References

1. Oerke EC, Dehne HW, Schönbeck F, Weber A (1994) Crop production and crop protection. Elsevier, Amsterdam
2. Wiese MV (1991) Compedium of wheat diseases. American Phytopathological Society, St Paul, MN, USA
3. Murray GM, Brennan JP (1998) Australasian Plant Pathol 27:212
4. Flor HH (1946) J Agric Sci 73:335
5. Keller M, Keller B, Schachermayr G, Winzeler M, Schmid JE, Stamp P, Messmer MM (1999) Theor Appl Genet 98:903
6. Mamluk OF (1998) Euphytica 100:45
7. Manninen AM, Laatikainen T, Holopainen T (1998) Trees 12:347
8. Monkiedje A, Spiteller M (2002) Biol Fertil Soils 35:393
9. Latteur G, Jansen JP (2002) Biocontrol 47:435
10. van de Veire M, Sterk G, van der Staaij M, Ramakers PMJ, Tirry L (2002) Biocontrol 47:101

11. Bliffeld M, Mundy J, Potrykus I, Fütterer J (1999) Theor Appl Genet 98:1079
12. Bieri S, Potrykus I, Fütterer J (2003) Mol Breed 11:37
13. Broglie K, Chet I, Holliday M, Cressman R, Biddle P, Knowlton S, Mauvais CJ, Broglie R (1991) Science 254:1194
14. Jach G, Gornhardt B, Mundy J, Logemann J, Pinsdorf E, Leah R, Schell J, Maas C (1995) Plant J 8:97
15. Lin W, Anuratha CS, Datta K, Potrykus I, Muthukrishnan S, Datta SK (1995) Biotechnology 13:686
16. Swords KMM, Liang J, Shah DM (1997) Genet Eng 19:1
17. Koltin Y, Day P (1976) Proc Natl Acad Sci USA 273:594
18. Kandel J, Koltin Y (1978) Exp Mycol 2:270
19. Koltin Y, Day PR (1975) Appl Microbiol 30:694
20. Puhalla JE (1968) Genetics 60:461
21. Field LJ, Bruenn JA, Chang TH, Pinhasi O, Koltin Y (1983) Nucleic Acids Res 11:2765
22. Koltin Y (1988) The killer system of *Ustilago maydis*: secreted polypeptides encoded by viruses. Marcel Dekker, New York
23. Gage MJ, Bruenn J, Fisher M, Sanders D, Smith TJ (2000) Mol Microbiol 41:775
24. Banuett F (1995) Annu Rev Genet 29:179
25. Bruenn J (2002) The double-stranded RNA viruses of *Ustilago maydis* and their killer toxins. In: Tavantzis SM (ed) dsRNA genetic elements: concepts and applications in agriculture, forestry, and medicine. CRC, Boca Raton
26. Winter W, Krebs H, Bänziger I (1995) Agrarforschung 2:325
27. Wilcoxson RR, Saari EE (1996) Bunt and smut disease of wheat: concepts and methods of disease management. Centro Internacional de Mejormamento de Maiz Y Trigo, Mexico, DF
28. McIntosh RA (1998) Euphytica 100:19
29. WTO: http://www.wto.org/english/tratop_e/dda_e/tnc_e.htm
30. Royer M (2005) http://www.aphis.usda.gov/ppq/ep/kb/
31. Fisher K, Schoen CC, Miedaner T (2002) Chancen der Resistenzzuechtung gegen Brandpilze bei Weizen fuer den oekologishen Pflanzenbau. Landessaatzuchtanstalt, Universität Hohenheim, Stuttgart-Hohenheim
32. Clausen M, Krauter R, Schachermayr G, Potrykus I, Sautter C (2000) Nat Biotechnol 18:446
33. Pellegrineschi A, Noguera LM, Skovmand B, Brito RM, Velazquez L, Salgado MM, Hernandez R, Warburton M, Hoisington D (2002) Genome 45:421
34. Schlaich T, Urbaniak BM, Malgras N, Ehler E, Birrer C, Meier L, Sautter C (2006) Plant Biotechnol J 4:63
35. Sautter C, Kräuter R, Schachermayr G (2000) Agrarforschung 7:545
36. Stadler MB, Stadler BM (2003) FASEB J 17:1141
37. Helm RM (2001) Joint FAO/WHO expert consultation on foods derived from biotechnology. Biotech 01/07
38. Smalla K, Heuer H, Gotz A, Niemeyer D, Krogerrecklenfort E, Tietze E (2000) Appl Environ Microbiol 66:4854
39. Ellstrand NC, Prentice HC, Hancock JF (1999) Annu Rev Ecol Syst 30:359
40. Pickett AA (1993) Hybrid wheat—results and problems. Advances in plant breeding. Paul Parey, Berlin
41. Waines JG, Hedge SG (2003) Crop Sci 43:451
42. Khan MN, Heyne EG, Arp AL (1973) Crop Sci 13:223
43. Weiss G (2001) Science 294:2067

Invited by: Professor Fiechter

Adv Biochem Engin/Biotechnol (2007) 107: 113–132
DOI 10.1007/10_2007_053

Published online: 31 March 2007

Recombinant DNA Technology in Apple

Cesare Gessler[1,2] (✉) · Andrea Patocchi[1,3]

[1]Plant Pathology, Institute of Integrative Biology, ETH Zürich, 8092 Zürich, Switzerland
Cesare.gessler@agrl.ethz.ch

[2]SafeCrop Centre c/o Istituto Agrario San Michele, Trento, Italy

[3]*Present address:*
Plant Pathology, Agroscope Changins-Wädenswil (ACW), Box 185, 8820 Wädenswil, Switzerland

Abstract This review summarizes the achievements of almost 20 years of recombinant DNA technology applied to apple, grouping the research results into the sections: developing the technology, insect resistance, fungal disease resistance, self-incompatibility, herbicide resistance, fire blight resistance, fruit ripening, allergens, rooting ability, and acceptance and risk assessment. The diseases fire blight, caused by *Erwinia amylovora*, and scab, caused by *Venturia inaequalis*, were and still are the prime targets. Shelf life improvement and rooting ability of rootstocks are also relevant research areas. The tools to create genetically modified apples of added value to producers, consumers, and the environment are now available.

Keywords Disease · *Malus* · Pest · Transgenic

1 Introduction

The genus *Malus* includes a variety of species all closely related and easily crossable. Agriculturally relevant is the species *Malus* × *domestica*, the edible apple. Peculiarities that are relevant to its use are self-incompatibility, leading to high heterozygosity, and the difference between apple as a crop and *Malus* in natural conditions. Commercial orchards are uniform as they generally consist of few varieties (cultivars) with each individual tree being the result of a rootstock and a grafted scion, both obtained by vegetative propagation. These original selections, made possible centuries earlier, were intensively replicated, and are currently present worldwide. In contrast, in natural conditions each tree is a result of a meiotic event and assembles characteristics from the mother tree and the pollen donor tree, so that in nature no two trees will ever be genetically equal. Vegetative propagation by grafting has allowed the selection of high-quality apples through the centuries, by propagating a single tree developed from a seed which coincidently had more appealing fruit and growth characteristics than other locally known varieties. Some of the popular cultivars, such as Golden Delicious, Granny Smith, or the more local Gravensteiner and Boskoop, are chance seedlings. Few cultivars are the result of an oriented breeding program. Although agronomically this genetic uniformity and the maintenance and distribution of a particular genotype is desirable, genetic uniformity is deadly for the resistance toward any natural enemy. In fact the most successful commercial apple cultivars have lost the efficacy of their resistance genes toward the most frequent fungal pathogens, first of all *Venturia inaequalis*, causal agent of scab, and secondly *Podosphaera leucotricha*, causal agent of apple mildew. Still efficacious resistance genes can be found in wild *Malus* species, but commercial apple cultivars need constant protection through fungicides, wherever the climatic conditions are favorable for either of these diseases. On average 10 to 15 fungicide applications are necessary during a season to produce scab-free fruits, with costs from US $500 to 1500 per hectare per year. Common commercial cultivars are also susceptible to the sporadic, yet highly damaging, fire blight infections caused by the bacterium *Erwinia amylovora*. Fire blight is in some parts of the world a constant threat. Control measures against fire blight are relatively inefficient; only the antibiotic streptomycin is reliable, but is banned in several countries.

Not only the high costs of controlling diseases, but also the increasing concern from consumers and environmentalists over pesticide residues in organic (copper residues) and conventional production have fostered the breeding of disease resistant cultivars. Therefore, most current apple breeding programs are oriented toward reducing the need for pesticides, without losing the high plant and fruit quality.

Even though the advantage of a resistant cultivar is evident, resistant cultivars do not yet dominate the market. The reason is simple: after each cross the seeds produced will lead to individuals differing from the mother, father, and each other; each seed will be genetically unique, and if selected will be a new cultivar with considerably different properties from those of the mother and father trees. This is a relevant limitation in apple breeding; each time we would like to introduce a new trait by breeding, a completely new genotype with new characteristics will be generated. Backcrossing with one of the ancestors is not possible (or only to a very limited extent) in apple breeding, and therefore the original cultivar with the incorporation of the trait of interest, mostly resistance to a pathogen, cannot be re-created. To eliminate unwanted genome segments a very time-consuming pseudo-backcrossing with different domestic cultivars is necessary. Even if modern DNA analysis methods, such as genetic maps, identification of DNA markers linked to traits of interest, and marker-assisted breeding, can help and accelerate the process, a new cultivar with fruits possessing a new taste, storage and conservation capacity, and a new tree form and growth patterns will be created. Last but not least this new cultivar also needs acceptance by the consumers, who often stick to a few favorite cultivars. Moreover, if not a single gene (single trait) but several have to be introduced, such as resistance to scab and mildew, to fire blight, or several genes for a single trait (e.g., several resistance genes against a single pathogen to achieve durable resistance), it becomes an almost impossible endeavor.

Under these premises, the introduction of specific gene(s) into a particular cultivar which already has all the necessary qualities, except the trait in question, is attractive. Recombinant DNA technology (gene technology) promises to do exactly this.

2 Developing the Technology

Apple was an early target for the emerging recombinant DNA technology. *Agrobacterium tumefaciens*, a disarmed Ti binary vector, and leaf fragments or callus cultures used in the original experiments were and still are the materials of choice [1]. The first transgenic apple plants contained a cassette with the genes for nopaline synthase and neomycin phosphotransferase (*npt*II, conferring resistance to the antibiotic kanamycin). The selection of the transformed cells, regeneration, and rooting were made in the presence of the antibiotic. The incorporation of the genes into the genome was confirmed by Southern blot analysis. The nopaline synthase gene continued to be expressed in greenhouse-grown plants several months after removal from the in vitro growth conditions. The transgenic apple plants showed a normal phenotype, except for a somewhat reduced capacity to root, which is an important trait if the transformant is a rootstock but irrelevant if the target is a scion cultivar.

application in nurseries during rootstock production. In fruit production there is little advantage, as herbicides damage trees only in exceptional situations and when normal safety precautions are not observed. The gene for resistance to the herbicide bialaphos (bilanafos) (bar, encoding phosphinothricin acetyltransferase) was introduced into *M. prunifolia* var. *ringo*. The herbicide resistance also served as the selectable marker. Transgenic plants were fully resistant to bialaphos [70]. The transformation and selection of the phosphinotricine (PPT, Basta) resistant apple clonal rootstock No. 545 offered some interesting information. PPT compared to *npt*II was inefficient as a selectable marker, giving a high frequency of escapes and a transformation frequency of 1.3% compared to 3.5–14% obtained with *npt*II and the same rootstock. However, once an appropriate line was selected, the plants were resistant to the commercially applied dosages of Basta [71].

7 Fire Blight Resistance

Probably the first and most important target of transgenic apple was fire blight resistance, which was pioneered by the Cornell University group led by Aldwinckle [72, 73]. As for the experiments performed to infer resistance against fungi and insects, several genes with different characteristics were used to transform apple cultivars and rootstocks with the aim of increasing their resistance to fire blight. Attacin A [74] and E genes were used to transform the rootstock cultivar Malling 26. In both in vitro and greenhouse trials the plants from the transformed line were more resistant. The GM plants supported a tenfold higher inoculum dosage before reaching the 50% lethal factor [75]. GM lines of the cultivar Galaxy, with the genes for attacin E and T4 lysozyme, have increased resistance to fire blight [76]. A set of German cultivars (Elstar, Pinova, Pilot, Pingo, Pirol, and Remo) have been similarly transformed with T4 lysozyme gene [77]. Unfortunately, we found no further report on the outcome of the mentioned fire blight resistance tests. Norelli et al. [78] overviewed the ongoing work and additionally reported transformation of Gala, Royal Gala, and Galaxy with the cecropin analogs SB 37 and Shiva 1, and a hen egg white lysozyme (HEWL). However, the best results were obtained with the attacin E. One particular line displayed only 5% shoot blight compared with approximately 60% of the untransformed Royal Gala control plants in 1998 field trials [79]. A short overview of the field trials with the various lytic protein transgenic lines was presented by Aldwinckle [72, 73].

Instead of using the constitutive promoter CaMV 35S, Liu et al. [80] genetically engineered into Royal Gala a modified cecropin SB37 gene (MB39), fused to a secretory coding sequence from barley alpha amylase and placed under the control of a wound-inducible osmotin promoter from tobacco.

Seven diploid transgenic lines were produced (transformation efficiency of 1.7%) of which three lines, all having multiple insertions, had a higher proportion of shoots with low infection scores than Royal Gala.

A similar approach was attempted by the German team of the Max Planck Institute in Lenburg and the Institute for Fruit Breeding in Dresden. The extracellular polysaccharide (EPS) amylovoran produced by *E. amylovora* is an essential pathogenicity and survival factor, building a capsule around the bacterium. *E. amylovora* bacteriophages produce an EPS depolymerase to degrade this capsule to finally reach the bacterium. The EPS depolymerase gene was inserted into an apple rootstock JTE H by *Agrobacterium*-mediated transformation and *npt*II as selection gene. In the first pathogenicity tests, four out of five transgenic plants were completely resistant to fire blight [81]. A further extension of the work used the cultivar Pinova as the host plant. In vitro, 61 out of 83 transgenic lines were significantly more resistant to fire blight. Some transgenic lines were transferred to greenhouse conditions and tested for increased resistance. The authors do not report any data, but state that the correlation between the in vitro data and the greenhouse data is a mere $r = 0.5$ [82]. The same gene has been used in pear and only two out of 15 lines showed consistent increase of fire blight resistance in vitro and in greenhouse tests. The two lines were also those with the highest EPS depolymerase expression [83].

The pathogen-induced plant resistance approach starts from the theory that the pathogen secretes substances which are recognized by the host and may initiate the defense cascade. It has been shown that by applying these substances prior to inoculation with a pathogen, an incompatible reaction can be induced. The *E. amylovora* effector protein, harpin, when sprayed on flowers protects against subsequent *E. amylovora* infection, probably by inducing the systemic acquired resistance responses. However, high dosages lead to cell death. The gene encoding for harpin N (*hrpN*) under the control of the promoter *Pgst1*, which is induced in *E. amylovora* challenged leaves [84, 85] and under the weak constitutive promoter *nos* with and without a signal peptide (SP) which should direct the harpin N to the intercellular space, were introduced into the rootstock M.26 [86]. Transgenic plants expressing *hrpN* under the control of the *nos* promoter have been obtained [86] as well as a few lines under the control of *Pgst1* [72]. In an overview Aldwinckle [73] reports that some lines showed an increase of resistance in the field. Also, the cultivar Galaxy has been transformed with the constructs *Pgst1*-SP-*hrpN*. Field trials for fire blight resistance are currently under way [73] and it will be highly interesting to know the outcome.

In all the above reported approaches to create a fire blight resistant plant, the resistance induction was entrusted to a non-*Malus* gene. Besides the harpin N, *E. amylovora* also produces the pathogenicity effector protein *dspE*, which interacts directly with four leucine-rich repeat (LRR) receptor-like serine/threonine kinases of apple (DIPM). If this interaction does not take place *E. amylovora* will not be able to infect that host, i.e., in the absence of the host DIPM receptor proteins the specific plant is a nonhost. Constructs with complete or partial hairpin *DIPM* gene sequences controlled by the CaMV 35S promoter were made. Several GM lines were recovered and in most the corresponding *DIPM* genes were silenced. However, no line with all four *DIPM* genes silenced was recovered. Preliminary fire blight resistance tests in growth chambers indicate that some lines are more resistant to artificial infection of the shoot tips [87]. As the authors point out, "resistance due to silencing of a native apple gene(s) is likely to be more acceptable to regulators, growers, and consumers than the addition of any foreign genes."

Overexpression of an apple own gene, linked to defense, was proposed by Aldwinckle. The *NPR1* protein in *Arabidopsis* is linked to the defense reaction and when overexpressed enhances resistance to *Pseudomonas syringae* and *Peronospora parasitica* [88]. Aldwinckle's group incorporated the apple ortholog *MpNR1* with the promoter Ppin2 (*E. amylovora* induced) into Galaxy and M.26. Two transgenic lines of Galaxy were clearly more resistant to fire blight, with 36% compared to 93% (controls) of the shoots blighted in a growth chamber test with potted plants [72].

8
Fruit Ripening

When discussing the pros and cons of replacing popular cultivars with high susceptibility to diseases with new cultivars possessing improved disease resistance, one of the principal arguments against any change is the lack of adequate storage and shelf life [89]. A number of genes have been identified which play a determinate role in the ripening and softening of fruit; however, ethylene remains a key factor. Ethylene is synthesized in plants from *S*-adenosyl methionine with the aid of two enzymes: 1-aminocyclopropane-1-carboxylic acid synthase (ACS), which converts *S*-adenosyl methionine to 1-aminocyclopropane carboxylic acid (ACC), and 1-aminocyclopropane carboxylic acid oxidase (ACO), which oxidizes ACC to ethylene.

Hrazdina et al. [90] transformed Royal Gala with an antisense sequence of one of the ACS genes. Transgenic lines were planted in the field and evaluated. GM plants developed normally and the fruits were, in most lines, similar to the nontransgenic controls. There was no significant decrease in ethylene production per fruit, nor in the other relevant parameters (soluble solids, pH, and penetration force) when the average value of the various parameters of all plants of a line was considered. However, the authors write that "on the percent basis ethylene production in apples from the individual trees from different transgenic lines was downregulated after 69 days of storage by 11 to 97% with the majority of fruits falling in the 60–80% range." Apparently the time course of ethylene presence was changed so that, at the last evaluation date, the fruit from the transgenic lines produced only half the amount of ethylene compared to the control fruits. The authors infer that the data indicate that the structural integrity of the fruit may be maintained simply under refrigerated storage and controlled humidity. Similar results were achieved by Galli et al. [91] with fruit from lines of transgenic Royal Gala with down-regulated ethylene synthesis. The fruits remained firm and had an increased resistance to shriveling, splitting, and spoilage following extended storage at room temperature.

In a very careful and accurate study the research group of the Department of Pomology, University of California, Davis, evaluated the impact of the suppression of ethylene on the fruit quality [92–94] using transgenic apple trees with ACC synthase or ACC oxidase enzyme activity suppression. Sucrose and fructose levels were lowered, malic acid degradation was reduced, and the volatile aroma ester and alcohol fractions were similarly reduced. In a collaborative work with Horticulture Research International, East Malling, England, some incongruence still remains, as the results suggest that sugar and acid composition are not under the direct control of ethylene and alcohol volatiles seem not to be influenced. However, for practical application, the firmness of the fruits from transgenic lines with suppressed ethylene biosynthesis remained almost constant after storage (shelf life), whereas in the control Greensleeves fruits it decreased dramatically [95]. These studies can be viewed from an application point of view but they also contribute highly to our understanding of the ripening processes in apple. Others do not have such an evident potential of application, but merit being mentioned as they contribute to basic knowledge. Cheng et al. [96] demonstrated the plasticity of the apple photosynthesis system by using antisense inhibition of sorbitol synthesis in GM apple; Kanamuaru et al. [97] were able to determine that S6PDH is a key enzyme regulating partitioning between sorbitol and sucrose in apple leaves.

Atkinson et al. [98] overexpressed polygalacturonase and obtained a range of new phenotypes, altering leaf morphology, plant water relation, stomata structure and function, as well as leaf attachment. Underexpression of polyphenol oxidase (PPO) (catechol oxidase), the enzyme responsible for enzymatic browning of apples, by use of an antisense PPO gene clearly led to reduced calli browning [99] and shoots had a similarly lower tendency of browning through the PPO activity [100].

9 Allergens

Malus has four allergenic proteins, Mald 1 to 4. They cross-react with the birch pollen Bet v 1 specific IgE antibodies, so that an apple allergy is common in patients with a birch pollen allergy. Apple cultivars have different allergenic potential, Golden Delicious being a cultivar with a high potential. A hairpin construct from Mald 1 was introgressed by *Agrobacterium*-mediated transformation with the selection gene *npt*II in the cultivar Elstar. Six plantlets displayed a significant reduction of Mald 1 production (at least tenfold) and induced significantly less reaction in patients than the control plantlets [101]. Specific RNAi silencing can therefore solve a diet problem for birch pollen allergenic patients.

10 Rooting Ability

Vegetative propagated plants depend on a high rooting capacity or have to be grafted on rooted "rootstocks". Clearly such a rootstock cultivar has to have, besides its specific growth characteristics (mostly vigor leading to dwarfing or allowing a high tree), a good rooting ability. Apple rootstocks are propagated by stool layering, seldom by rooting of cuttings as some are recalcitrant to root from cuttings even with the use of auxin. Root-inducing genes have been characterized in *Agrobacterium rhizgenes* (*rol*A, *rol*B, and *rol*C), and contribute to causing "hairy root" disease in the host. The bacterium introduces parts of its DNA containing the *rol* genes into the host plant, which reacts by producing additional roots and often assuming a dwarfing stance. Almost all transgenic works with the *rol* genes have their origin at the Department of Horticulture at the Swedish University of Agricultural Science in Alnarp. Transforming the apple rootstock M.26 with *rol*A controlled by its own promoter resulted in four lines with variable growth reduction and wrinkled leaves [102]. Incorporation of the *rol*B gene, also under the control of its own promoter, into the rootstock M.26 increased auxin sensitivity and rooting ability [103]. To think of commercialization of a rootstock with increased ability to root, it is necessary to demonstrate that no negative effect will be transmitted to the graft scion. Zhu and Welander [104] used the cultivar Gravenstein as a test scion. Its growth characteristics were not influenced under unlimited nutrient conditions. The specific root length of the M.26 rootstock was significantly reduced, from which it can be speculated that under limited nutrient supply, as in orchard conditions, it may induce tree dwarfing.

M9 is another very popular rootstock with excellent dwarfing characteristics; however, it roots badly from cuttings. GM M9 with the *rol*B gene roots extremely well. In vitro rooting ability went from almost nil to almost 100% with 3.5 to 9.5 roots per cutting [105]. Also, the rootstock Jork 9 does not root readily from cuttings. Incorporation of *rol*B dramatically increased the rooting of shoots in the absence of externally added auxin. The control untransformed shoots were able to reach almost the same rooting ability with the addition of indole-3-butyric acid (IBA) [106]. The rootstock Jork 9 and the various transgenic lines have been used to study plant response to the rooting process, natural or auxin induced [107]. Welander et al. [108] report that "the permission for field trial on the transformed rootstocks has been obtained from the Board of Agriculture in Sweden." Plants of different transformed clones and the untransformed controls of M.26 and M9/29 have been produced and grown in the field for 2 years. Five apple cultivars commonly used in Sweden and Europe have been budded onto the rootstocks to evaluate the influence of transgenes on the growth and development of the grafted cultivars.

The *rol*C gene has been introduced into a Japanese rootstock (*M. prunifolia* var. *ringo* Asami Mo 84 A), which roots well but has not the desired dwarfing ability. The aim was to determine whether *rol*C can reduce growth characteristics without altering the rooting characteristics. Some transformants had shortened internodes, some reduced height, some both and some were normal; rooting was intensive in some transformants. The authors state that a few lines may be suitable candidates for dwarfing rootstocks [109]. Radchuk and Korkhovoy [110] transformed the scion cultivar Florina with *rol*B with the theory that multiplying and growing it on its own roots would reduce costs and accelerate production of plant material. They obtained various lines with enhanced rooting ability. From the data for 2 years of greenhouse experiments they report no change to the above-ground growing characteristics compared to those of the original untransformed Florina.

11
Acceptance and Risk Assessment

Apple is vegetatively propagated and seeds only have a role in breeding. Transformation of the apple into a weed through a selective advantage gained from a foreign gene is not plausible under any imagination. Environmental risk is restricted to the gene products and is not inherent to apple. Apple is a fresh product often consumed raw, and therefore consumers are particularly attentive to any manipulation. Currently it is improbable that foreign genes will be acceptable under European law where transgenics are highly regulated and have to be declared as such. So the arguments listed in 1996 by Koller et al. [111, 112], leading to the opinion that transgenic apple carrying foreign DNA will not be commercialized in the near future, are still valid.

Almost all work cited in this review, excluding the few specially mentioned exceptions, relied on the selection gene *npt*II, on non-apple gene promoters (CaMV 35S amongst others), and on *Agrobacterium tumefaciens*-mediated transformation. Often, for experimental purposes, genes not influencing the target trait were used, such as the gene producing β-glucuronidase (GUS). To our knowledge no environmental risk studies specific to apple have been published. The only argument studied was the possibility that particular insects may become resistant to the Bt toxin [113]. The development of Bt resistance was analyzed using simulation experiments in the agrosystems apple and clover, both hosts for various leafrollers [114]. Probably the researchers are still concerned with producing acceptable GM apple cultivars with a commercial interest [115] and possibly having environmental benefits, such as reduction of pesticide use.

However, an in-depth discussion should be conducted on any gene used. For example, genes expressing the biotin binding proteins avidin and strepavidin, as indicated above, and used to infer insect resistance [44] in fruits would also bind biotin for the consumer and may lead to reduced vitamin B absorption similar to a surplus of egg consumption.

What is acceptable, however, is debated and some argue that only apple-derived genes will have the possibility of encountering the favor of the producer and consumer. In this context the research by Kassardjian et al. [116] on purchase behavior is interesting, mainly as a guideline that includes innovative ideas on how to elicit consumers' willingness to purchase GM apples. The goal of their research was to try a new methodology (thought-listing technique and questionnaires) to elicit consumer willingness to pay for GM food in New Zealand. However, it is not that the researchers are inactive; in popular journals [117–119], interviews, and web sites arguments for and against the use of GM apples are presented and debated.

12 Conclusion

In the next few years many "apple own" resistance genes will be sequenced and transferred into some test cultivars, probably under the control of their own promoters. RNA interference technology will be able to block some unwanted traits. Pathogen-derived genes inducing host resistance will be available. Apple own promoters expressing genes only where and when desired will no longer be a mirage but will appear slowly on our horizon. With the "clean vector technology" [36] allowing the removal of the selection marker, it will be possible to produce cisgenic apples, i.e., apple plants which will be modified exclusively with *Malus* genes and controlled by *Malus* promoters. Many of the discussions about risks, phantom or truly demonstrated, will be obsolete. However, the insertion site will not correspond to the original site of the gene and this could lead to epigenetic effects. Major effects are readily discovered and such lines are discharged; subtle effects are more difficult to discover, so we have to strive to devise smart experiments allowing for selection of lines which harbor no surprises.

The apple cropping system based on artificial vegetative multiplication of the particular genotype and its planting over large areas has rendered this crop susceptible [120]. The possibility offered by DNA recombinant technology can be used to replace nonfunctional resistance alleles by not-overcome alleles (gene therapy). The benefits of GM apple resistant to various diseases will be real, not only for the owner of a patent but also for the producer, environment, and consumer. The reduction of fungicide and, in some parts of the world, antibiotic use alone justifies all the efforts [121]. It remains to be seen how long it will take until a broad acceptance by the public is achieved.

References

1. James DJ, Passey AJ, Barbara DJ, Bevan M (1989) Plant Cell Rep 7:658
2. Maheswaran G, Welander M, Hutchinson JF, Graham MW, Richards D (1992) J Plant Physiol 139:560
3. Cheng JS, Dandekar AM, Uratsu SL (1992) Acta Hortic Sin 19:101
4. Lambert C, Tepfer D (1992) Theor Appl Genet 85:105
5. Norelli JL, Aldwinckle HS (1993) J Am Soc Hortic Sci 118:311
6. Trifonova A, Savova D, Ivanova K (1994) *Agrobacterium*-mediated transformation of the apple cultivar Granny Smith. In: Schmidt H, Kellerhals M (eds) Progress in temperate fruit breeding. Kluwer, Dordrecht, p 343
7. Yao JL, Cohen D, Atkinson R, Richardson K, Morris B (1995) Plant Cell Rep 14:407
8. Schaart JG, Puite KJ, Kolova L, Pogrebnyak N (1995) Euphytica 85:131
9. Yao JL, Cohen D, Atkinson R, Richardson K, Morris B (1995) Plant Cell Rep 14:407
10. Dolgov SV, Miroshnichenko DN, Schestibratov KA (2000) Acta Hortic 538:619
11. Yepes LM, Aldwinckle HS (1994) Plant Cell Tissue Organ Cult 37:257
12. Oraguzie NC, Rikkerink E, Gardiner S, Bus V, Currie A, Rusholme R, Volz R (2004) Recent Res Dev Genet Breed 1(II):223
13. Reim S, Hanke V (2004) Acta Hortic 663:419
14. James DJ, Passey AJ, Baker SA (1994) Euphytica 77:119
15. James DJ, Passey AJ, Baker SA, Wilson FM (1996) Biotechnology 14:56
16. Rühmann S, Teutter D, Fritsche S, Briviba K, Szankowski I (2006) J Agric Food Chem 54:4633
17. de Bondt A, Eggermont K, Penninckx I, Goderis I, Browkaert WF (1996) Plant Cell Rep 15:549
18. Zhang Z, Jing S, Wang G-L, Fang H-J, Wu L-P (1997) Acta Hortic Sin 24:378
19. Puite KJ, Schaart JG (1996) Plant Sci 119:125
20. Puite K, Schaart J (1999) Acta Hortic 484:547–553
21. Bolar JP, Brown SK, Norelli JL, Aldwinckle HS (1999) Plant Cell Tissue Organ Cult 55:31
22. Song KJ, Park SW, Ahn SY, Hwang JH, Shin YU (2001) Acta Hortic 560:211
23. McAdam O, Connell D, Mac An Tsaoir S, Copeland R (2004) Acta Hortic 663:483
24. Wilson FM, James DJ (2003) J Hortic Sci Biotechnol 78:656
25. Welander M, Maheswaran G (1992) J Plant Physiol 140:223
26. Yamashita H, Daimon H, Akasaka-Kennedy Y, Masuda T (2004) J Jpn Soc Hortic Sci 73:505
27. Pawlicki-Jullian N, Sedira M, Welander M (2002) Plant Cell Tissue Organ Cult 70:163
28. Norelli J, Mills JA, Aldwinckle H (1996) HortScience 31:1026
29. ALiu Q, Salih S, Hammerschlag F (1998) Plant Cell Rep 18:32
30. Maddumage R, Fung RMW, Weir I, Ding H, Simons JL, Allan AC (2002) Plant Cell Tissue Organ Cult 70:77

31. Gittins JR, Hiles ER, Pellny TK, Biricolti S, James DJ (2001) Mol Breed 7:51
32. Gittins JR, Pellny TK, Hiles ER, Rosa C, Biricolti S, James DJ (2000) Planta 210: 232
33. Gittins JR, Pellny TK, Biricolti S, Hiles ER, Passey AJ, James DJ (2003) Transgenic Res 12:391
34. Flachowsky H, Birk T, Hanke V (2004) Acta Hortic 663:425
35. Zhu LH, Li XY, Ahlman A, Xue ZT, Welander M (2004) Acta Hortic 663:503
36. Krens FA, Pelgrom KTB, Schaart JG, de Nijs APM, Rouwendal GJA (2004) Acta Hortic 651:101
37. Rat-Morris E, Lespinasse Y (1995) Phytoma 471:15
38. Qubbaj T, Reineke A, Zebitz CPW (2005) Entomol Exp Appl 115:145
39. Cevik V, King GJ (2002) Genome 45:939
40. James DJ, Passey AJ, Easterbrook MA, Solomon MG, Barbara DJ (1992) Phytoparasitica 20[suppl]:83
41. James DJ, Passey AJ, Webster AD, Barbara DJ, Dandekar AM, Uratsu SL, Viss P (1993) Acta Hortic 336:179
42. Dandekar AM, McGranahan GH, Uratsu SL, Leslie C, Vail PV, Tebbets JS, Hoffmann D, Driver J, Viss P, James DJ (1992) In: Haskell PT (ed) Pests and diseases. BCPC, Farnham, p 741
43. Cheng JS, Li WG, Meng XM, Tian YC, Mang KQ (1998) China Fruits 4:21
44. Markwick NP, Docherty LC, Phung MM, Lester MT, Murray C, Yao JL, Mitra DS, Cohen D, Beuning LL, Kutty Amma S (2003) Transgenic Res 12:671
45. Wong KW, Harman GE, Norelli JL, Gustafson HL, Aldwinckle HS (1998) Acta Hortic 484:595
46. Bolar JP, Norelli JL, Harman GE, Brown SK, Aldwinckle HS (1999) In: Altman A, Ziv M, Izhar S (eds) Plant biotechnology and in vitro biology in the 21st century. Kluwer, Dordrecht, p 465
47. Bolar JP, Norelli JL, Wong KW, Hayes CK, Harman GE, Aldwinckle HS (2000) Phytopathology 90:72–77
48. Norelli JL, Bolar JP, Harman GE, Aldwinckle HS (2000) Acta Hortic 538:617
49. Bolar JP, Norelli JL, Harman GE, Brown SK, Aldwinckle HS (2001) Transgenic Res 10:533
50. Aldwinckle HS, Norelli JL, Bolar JP, Ko KS, Harman GE, Brown SK (1999) In: Altman A, Ziv M, Izhar S (eds) Plant biotechnology and in vitro biology in the 21st century. Kluwer, Dordrecht, p 449
51. Hanke V, Hiller I, Klotzsche G, Winkler K, Egerer J, Richter K, Norelli JL, Aldwinckle HS (2000) Acta Hortic 538:611
52. deBondt A, Zaman S, Broekaert W, Cammune B, Keulemans J (1998) Acta Hortic 484:565
53. Broothaerts W, deCubber K, Zaman S, Coppens S, Keulemans J (2000) Acta Hortic 521:91
54. deCubber K, Broothaerts W, Lenaerts T, Keulemans J (2000) Acta Hortic 525:309
55. Faize M, Malnoy M, Dupuis F, Chevalier M, Parisi L, Chevreau E (2003) Phytopathology 93:1496
56. Chevreau E, Faize M, Dupuis F, Sourice S, Parisi L (2004) Acta Hortic 663:447
57. Krishnamurthy K, Balconi C, Sherwood J, Giroux M (2001) Mol Plant Microbe Interact 14:1255
58. Chevreau E, Dupuis F, Ortolan C, Parisi L (2001) Acta Hortic 560:323
59. Faize M, Sourice S, Dupuis F, Parisi L, Gautier MF, Chevreau E (2004) Plant Sci 167:347

60. Sansavini S, Barbieri M, Belfanti E, Tartarini S, Vinatzer B, Gessler C, Silfverberg E, Gianfranceschi L, Hermann D, Patocchi A (2004) Riv Frutticolt Ortofloricolt 66:54
61. Vinatzer BA, Patocchi A, Gianfranceschi L, Tartarini S, Zhang H-B, Gessler C, Sansavini S (2001) Mol Plant Microbe Interact 14:508
62. Barbieri M, Belfanti E, Tartarini S, Vinatzer BA, Sansavini S, Silfverberg-Dilworth E, Gianfranceschi L, Hermann D, Patocchi A, Gessler C (2003) HortScience 38:329
63. Sansavini S, Barbieri M, Belfanti E, Tartarini S, Vinatzer B, Gessler C, Silfverberg E, Gianfranceschi L, Hermann D, Patocchi A (2003) Acta Hortic 622:113
64. Belfanti E, Barbieri M, Tartarini S, Vinatzer B, Gennari F, Paris R, Sansavini S, Silfverberg-Dilworth E, Patocchi A, Hermann D, Gianfranceschi L, Gessler C (2004) Acta Hortic 663:453
65. Belfanti E, Silfverberg-Dilworth E, Tartarini S, Patocchi A, Barbieri M, Zhu J, Vinatzer BA, Gianfranceschi L, Gessler C, Sansavini S (2004) Proc Natl Acad Sci USA 101:886
66. Silfverberg-Dilworth E, Patocchi A, Belfanti E, Tartarini S, Sansavini S, Gessler C (2005) Abstract Plant and Animal Genome XIII, January 15–19 2005, San Diego, CA, USA
67. Silfverberg-Dilworth E, Besse S, Paris R, Belfanti E, Tartarini S, Sansavini S, Patocchi A, Gessler C (2005) Theor Appl Genet 110:1119
68. van Nerum I, Incerti F, Keulemans J, Broothaerts W (2000) Acta Hortic 538:625
69. Broothaerts W, Keulemans J, van Nerum I (2004) Plant Cell Rep 22:497
70. Ogasawara H, Ueda S, Harada T, Ishikawa R, Niizeki M, Saito KI (1994) Bull Fac Agric Hirosaki 57:1
71. Dolgov SV, Skryabin KG (2004) Acta Hortic 663:499
72. Aldwinckle HS, Borejsza Wysocka EE, Malnoy M, Brown SK, Norelli JL, Beer SV, Meng X, He SY, Jin QL (2003) Acta Hortic 622:105
73. Aldwinckle HS, Norelli JL, Borejsza Wysocka E, Reynoird JP (2000) IOBC WPRS Bull 23:123
74. Norelli J, Aldwinckle H, Destefano Beltran L, Jaynes J (1993) Acta Hortic 338:385
75. Norelli JL, Aldwinckle HS, Destefano BL, Jaynes JM (1994) Euphytica 77:123
76. Ko K, Norelli JL, Brown SK, Aldwinckle HS, During K (1999) In: Altman A, Ziv M, Izhar S (eds) Plant biotechnology and in vitro biology in the 21st century. Kluwer, Dordrecht, p 507
77. Hanke V, During K, Norelli JL, Aldwinckle HS (1999) Acta Hortic 489:253
78. Norelli JL, Borejsza Wysocka EE, Momol MT, Mills JZ, Grethel A, Aldwinckle HS, Ko K, Brown SK, Bauer DW, Beer SV, Abdul-Kader AM, Hanke V (1999) Acta Hortic 489:295
79. Norelli JL, Mills JAZ, Momol MT, Aldwinckle H (1999) Acta Hortic 489:273
80. Liu Q, Ingersoll J, Owens L, Salih S, Meng R, Hammerschlag F (2001) Plant Cell Rep 20:306
81. Sule S, Kiss E, Kim WS, Geider K (2002) Acta Hortic 590:407
82. Hanke V, Kim WS, Geider K (2002) Acta Hortic 590:393
83. Malnoy M, Faize M, Venisse JS, Geider K, Chevreau E (2005) Plant Cell Rep 23:632
84. Reynoird JP, Abdul-Kadar AM, Bauer SW, Borejsza Wysocka E, Norelli JL, Aldwinckle HS (2000) Plant Mol Biol Rep Suppl 18:22
85. Malnoy M, Reynoird JP, Borejsza Wysocka E, Aldwinckle HS (2006) Transgen Res 15:83
86. Abdul-Kader AM, Norelli JL, Aldwinckle HS, Bauer DW, Beer SV (1999) Acta Hortic 489:247
87. Borejsza Wysocka EE, Malnoy M, Meng X, Bonasera JM, Nissinen RM, Kim JF, Beer SV, Aldwinckle HS (2004) Acta Hortic 663:469
88. Cao H, Li X, Dong X (1998) Proc Natl Acad Sci USA 95:6531

89. James D, Passey A, Baker S, Wilson F, Stow J, Colgan R, Hiles E, Massiah A, Vaughan S, Blakesley D, Simpson D, Sargent D, Bulley S, Hedden P, Phillips A, Biricolti S, Mazzara M, Uratsu S, Labavitch J, Dandekar AM (2003) Acta Hortic 622:97
90. Hrazdina G, Kiss E, Rosenfield C, Norelli JL, Aldwinckle HS (2000) In: Hrazdina G (ed) Use of agriculturally important genes in biotechnology. IOS, Amsterdam, p 26
91. Galli Z, Kiss E, Hrazdina G, Heszky L (2003) Int J Hortic Sci 9:65
92. Defilippi BG, Dandekar AM, Kader AA (2004) J Agric Food Chem 52:5694
93. Defilippi BG, Dandekar AM, Kader AA (2005) J Agric Food Chem 53:3133
94. Defilippi BG, Kader AA, Dandekar AM (2005) Plant Sci 168:1199
95. Dandekar AM, Teo G, Defilippi BG, Uratsu SL, Passey AJ, Kader AA, Stow JR, Colgan RJ, James DJ (2004) Transgenic Res 13:373
96. Cheng LL, Zhou R, Reidel EJ, Sharkey TD, Dandekar AM (2005) Planta 220:767
97. Kanamaru N, Ito Y, Komori S, Saito M, Kato H, Takahashi S, Omura M, Soejima J, Shiratake K, Yamada K, Yamaki S (2004) Plant Sci 167:55
98. Atkinson RG, Schroder R, Hallett IC, Cohen D, MacRae EA (2002) Plant Physiol 129:122
99. Murata M, Nishimura M, Murai N, Haruta M, Homma S, Itoh Y (2002) Biosci Biotech Biochem 2001 65:383
100. Murata M, Haruta M, Murai N, Tanikawa N, Nishimura M, Homma S, Itoh Y (2000) J Agric Food Chem 48:5243
101. Gilissen LJWJ, Bolhaar STHP, Matos CI, Rouwendal GJA, Boone MJ, Krens FA, Zuidmeer L, van Leeuwen A, Akkerdaas J, Hoffmann Sommergruber K, Knulst AC, Bosch D, van de Weg WE, van Ree R (2005) J Allergy Clin Immunol 115:364
102. Holefors A, Xue ZT, Welander M (1998) Plant Sci 136:69
103. Welander M, Pawlicki N, Holefors A, Wilson F (1998) J Plant Physiol 153:371
104. Zhu LH, Welander M (1999) Plant Sci 147:75
105. Zhu LH, Holefors A, Ahlman A, Xue ZT, Welander M (2001) Plant Sci 160:433
106. Sedira M, Holefors A, Welander M (2001) Plant Cell Rep 20:517
107. Sedira M, Butler E, Gallagher T, Welander M (2005) Plant Sci 168:1193
108. Welander M, Zhu LH, Li XY (2004) Acta Hortic 663:437
109. Igarashi M, Ogasawara H, Hatsuyama Y, Saito A, Suzuki M (2002) Plant Sci 163:463
110. Radchuk VV, Korkhovoy VI (2005) Plant Cell Tissue Organ Cult 81:203
111. Koller B, Gessler C, Bertschinger L, Kellerhals M (1996) In: Schulte E, Käppeli O (eds) Gentechnisch veränderte Krankheits und Schädlingsresistente Nutzpflanzen. Eine Option für die Landwirtschaft? The Swiss National Foundation for Scientific Research, Bern, p 297
112. Koller B, Gessler C, Bertschinger L, Kellerhals M (1997) IOBC/WPRS Bull 20:150
113. Wearing CH, Hokkanen HMT (1994) Biocontrol Sci Technol 4:573
114. Caprio MA, Suckling DM (2000) J Econ Entomol 93:173.
115. James D, Passey A, Baker S, Wilson F, Stow J, Colgan R, Hiles E, Massiah A, Vaughan S, Blakesley D, Simpson D, Sargent D, Bulley S, Hedden P, Phillips A, Biricolti S, Mazzara M, Uratsu S, Labavitch J, Dandekar AM (2003) Acta Hortic 622:97
116. Kassardjian E, Gamble J, Gunson A, Jaeger SR (2005) Brit Food J 107:541
117. Vanloqueren G, Baret PV (2005) Fruit Belge 73:117
118. Dandekar AM, Fisk HJ, McGranahan GH, Uratsu SL, Bains H, Leslie CA, Tamura M, Escobar M, Labavitch J, Grieve C, Gradziel T, Vail PV, Tebbets SJ, Sassa H, Tao R, Viss W, Driver J, James D, Passey A, Teo G (2002) HortScience 37:281
119. Bertschinger L, Kellerhals M, Theiler R, Frey J, Gafner J, Gessler C (2000) Obst Weinbau 136:363
120. MacHardy WE, Gadoury DM, Gessler C (2001) Plant Dis 85:1036
121. Norelli JL, Jones AL, Aldwinckle HS (2003) Plant Dis 87:756

Invited by: Professor Sautter

Adv Biochem Engin/Biotechnol (2007) 107: 133–151
DOI 10.1007/10_2007_056

Published online: 31 March 2007

Prospects for Biopolymer Production in Plants

Jan B. van Beilen · Yves Poirier (✉)

Département de Biologie Moléculaire Végétale, Université de Lausanne, Bâtiment Biophore, 1015 Lausanne, Switzerland
yves.poirier@unil.ch

Abstract It is likely that during this century polymers based on renewable materials will gradually replace industrial polymers based on petrochemicals. This chapter gives an overview of the current status of research on plant biopolymers that are used as a material in non-food applications. We cover technical and scientific bottlenecks in the production of novel or improved materials, and the potential of using transgenic or alternative crops in overcoming these bottlenecks. Four classes of biopolymers will be discussed: starch, proteins, natural rubber, and poly-β-hydroxyalkanoates. Renewable polymers produced by chemical polymerization of monomers derived from sugars, vegetable oil, or proteins, are not considered here.

Keywords Bioplastics · Rubber · Silk · Starch · Polyhydroxyalkanoate

Abbreviations
mclPHA Medium-chain length poly-β-hydroxyalkanoate
PHA Poly-β-hydroxyalkanoate

PHB	Poly-β-hydroxybutyrate
SALB	South American leaf blight
t/year	Tons per year
TPS	Thermoplastic starch

1 Introduction

It is nowadays almost taken for granted that plastics are made from mineral oil. However, this has not always been the case, as one of the first plastics, called collodion, was made from cotton cellulose in the mid-nineteenth century. The body of a 1941 Ford demonstration vehicle consisted of plant fibers, soy protein polymers, and rubber tires made from the plant Goldenrod. The low cost and reliable supply of fossil fuels put an end to that. In less than 20 years, petroleum-derived plastics almost completely replaced plant-based materials. Now, the pendulum appears to be swinging back. Three factors are of importance in the resurgence of the use biopolymers with plastic or elastomeric properties (referred to as bioplastics) in industrial and consumer products: economics, public acceptance, and regulation. Price and properties already allow some bioplastics to successfully compete with petrochemical plastics, due in part to a strong increase in oil price. For most bioplastics, however, significant efforts on the raw materials side and on processing technology are still required to make them competitive. Public acceptance of non-food applications of agricultural products is not a big issue. However, public acceptance of the use of transgenic plants for non-food purposes is not assured, especially in the European Union [1]. Life cycle assessments and a mounting concern over climate change due to greenhouse gas production may exert increasing regulatory pressure to shift from petrochemicals to renewable materials (e.g., by CO_2 taxation). In general, production of plastics from plant biopolymers promises to offer the potential of reliable (domestic) supplies, jobs in rural communities, sustainable production, lower greenhouse gas production, and competitive prices.

2 Routes to Biopolymers: White vs. Green Biotechnology

2.1 Biopolymers from Bacterial Fermentation

White or industrial biotechnology also provides routes for obtaining biopolymers. Plant biomass can be converted to glucose, fatty acids, or other small molecules, either as the main product or as a waste stream from other production processes. These small compounds may then be converted to bioplastics via microbial fermentation or chemical polymerization [2]. For example, poly-β-hydroxyalkanoates (PHA), biocellulose, xanthan, silk, and polythioesters, can be produced by recombinant or wild-type microorganisms in fermentation processes [3], while polylactic acid, polycaprolactone, and other (partially renewable) polyesters, such as polytrimethylene terephtalate (e.g., Sorona by Dupont) and polybutylene succinate (e.g., Bionolle by Showa), are produced using chemical polymerization of substrates that are, at least in part, renewable and generated by fermentation [2]. It is likely that the production processes of bioplastics will be part of future biorefineries, which are now in an early stage of development (with the exception of starch and paper mills). This early stage implies that in biorefineries the processing costs still determine the economic viability of bioproducts, and a great potential for streamlining and improved process integration exists. As biorefineries mature, the focus will shift away from processing to the raw materials, as has happened in the petrochemical industry.

2.2 Why Biopolymer Production in Plants?

Prices of raw materials for the production of bioplastics using white biotechnology are already on the rise. For example, sugar prices have become very volatile due to the strong demand. In the USA alone, 50×10^6 t of corn will be used in 2006 to produce ethanol fuel. At the same time, world production of grains in 2006 is expected to fall short of consumption by 60×10^6 t on a total of approximately 2×10^9 t, leading to upward price pressure. Similarly, increasing amounts of vegetable oils are being converted to biodiesel. Again, the current world production of vegetable oils is barely higher than the demand for food applications. In other words, it is likely that the price of sugars and vegetable oils (and energy-rich waste streams) will become tightly linked to the price of oil. Direct production of polymers in plants may circumvent this price issue to some extent.

Another factor to consider is the life cycle assessment: does the production of bioplastics really consume less raw material and energy, and produce less CO_2, than the production of petroleum-based plastics? Some analyses have shown that if the amount of polymer in plant material is high enough and the remaining biomass is used to generate energy for polymer processing, the life cycle assessments favor plant biopolymers over petrochemical plastics [4]. Life cycle assessments also favor the production of biobased polymers over biofuels if land-use is also included in the analysis [5]. However, new methods could also provide highly efficient chemical routes to polymers, such as for poly-β-hydroxybutyrate (PHB) [6].

2.3
The Role of Transgenic Plants in the Production of Biopolymers

What is the potential role of transgenic plants in the production of novel or improved biopolymers from crops? Simple targets are increased amounts of the desired biopolymer relative to other plant components (e.g., starch content), or decreased amounts of other compounds in the plant that interfere with processing (e.g., lignin, proteins, pectin, hemicellulose). More complicated research targets can be considered as well. Genes may be altered or introduced to change the substrate range or processivity of polymerases, the structure and amount of precursors available for polymerization may be adjusted, new genes may be introduced in plants to obtain polymers with different properties (e.g., by changing the ionic charge, composition, chemical reactivity, stability, solubility, melting and other thermoplastic properties), and gene regulation may be altered. Potentially, much can be accomplished. However, do these changes improve the economics of biopolymer production in crop plants? Any polymer modification could lead to adverse effects, such as lowered concentrations of the polymer, increased difficulty in processing, reduced plant growth and seed germination, or other undesired effects. Other aspects of the application of plant biotechnology for biopolymer production must also be taken into account, such as crop identity preservation, gene transfer to other non-crop plants, limited flexibility of production in plants versus bacteria, length of time to market for transgenic plants, transgene methods versus fast-track breeding and tilling, the use of non-food plants to avoid controversy or litigation, the cost and time frame of registering and patenting transgenic crops, low marginal costs for established technologies, and the role of the technology development time gap. In other words, demonstrating the technical feasibility of producing a modified polymer is probably the easiest step, but bringing the transgenic plant to market entails many difficult and expensive steps.

Important key questions that should be addressed at an early stage in the development of crop-based bioplastics are:

1. Is it possible to gain sufficient control over the properties of biopolymers in planta compared to the relative ease of control over composition and properties in chemical polymerization and in fermentation?
2. Is the significant investment of creating transgenic plants or plant breeding for bioplastic production justified by the economic value of the product, in view of the current main use of bioplastics in low-cost applications?

3
Plant Biopolymers

In this chapter, we have singled out four specific (classes of) biopolymers for special attention: starch, natural rubber, protein-based polymers, and PHA.

Starch, natural rubber, and proteins such as zein, gluten, and soy-protein, are naturally synthesized by plants. For these products, polymer productivity and quality may be optimized by plant breeding, targeted genetic changes, and improved processing technology. In contrast, PHA, fibrous proteins such as silk and elastin, and non-ribosomally synthesized proteins are not naturally synthesized in plants but may be produced in transgenic plants.

Cellulose, hemicellulose, and lignin are other major plant biopolymers, which will not be considered here. Only cellulose has important applications in its unmodified (e.g., cotton, fibers, and wood), and modified (e.g., cellulose acetate, which is produced at about 750 000 t/year) forms. Plants produce many other biopolymers that presently have relatively few applications in the non-food sector (Table 1). This could change if larger amounts of the materials become available, for example as co-products of biofuel production. Other biopolymers that are now isolated from fungi or bacteria could also be produced in plants.

3.1 Starch

Starch is the second major agricultural commodity after cellulose, is the least expensive food commodity, and has numerous industrial applications. It is the cheapest and easiest to handle biopolymer. Due to its abundance and low price (world production is 57×10^6 t/year, at around 0.30 €/kg depending on the source of the starch), it has found numerous applications in the non-food sector. To give a size perspective, non-food uses of starch in EU15 (the 15 member states of the EU before May 2004) amounted to 3.6×10^6 t/year, or about 13% of the total starch market in these countries [7].

Currently, only about 40 000 t/year are converted to plastic materials by a range of small and large companies worldwide (Table 2). Most of this bioplastic is marketed as biodegradable, and is used for packaging films and foams, and for disposables (e.g., cups and plates, plant pots, and bags). The growth potential of this market is high, with many studies referring to future market sizes in the range of 1×10^6 t/year [8].

Due to the importance of starch, its biosynthesis and ways to modify its properties have been studied in depth. A great effort has gone into genetic modifications affecting starch biosynthesis in plants. The so-called starch-enhancement technology has increased the amount of starch relative to the other components in potato, yielding more starch per hectare and lower processing costs (reviewed in [9]). Efforts to change the properties of starch in planta have focused on the ratio between amylose and amylopectin, the branching pattern of amylopectin, synthesis of phosphate-substituted starches, and the production of starches from new crops [10–12]. Until now, only the high-amylose and amylose-free starches have been commercialized. A major conclusion of research in this field is that the effects of a single ge-

Table 1 Selection of plant biopolymers and their applications

	Chemical structure and source	Applications as material
Cellulose	Polysaccharide: 1,4-linked β-D-glucose, most abundant component of terrestrial biomass. Can be derivatized to ethers and esters (with acetate, propionate, butyrate, etc.)	Nitrocellulose, cellophane,carboxy-methylcellulose, Tencel fiber, cellulose acetate
Hemicellulose	Polysaccharides: xylan, glucuronoxylan, arabinoxylan, glucomannan, and xyloglucan, present in almost all cell walls along with cellulose	Limited use as source of chemicals
Lignin	Complex (irregular) polyphenolic macro-molecule making up a quarter to a third of the dry mass of wood	Limited use as polymer and source of chemicals
Pectin	Various polysaccharides containing 1,4-linked α-D-galacturonic acid units, and L-rhamno-pyranose units, linear and branched molecules.	Edible films
Inulin	Polysaccharide: linear β $(2 \rightarrow 1)$-linked fruct-ose chains attached to a sucrose molecule. Belongs to fructan-group: alternative storage carbohydrate in the vacuole of ~ 15% of flowering plant species.	Mainly used to produce inulin syrup. Carboxy-methyl inulin is used as antiscalant.
Cutin	Polyester found on the surface of plants	None
Suberin	Complex (irregular) biopolymer consisting of ω-hydroxyalkanoates, di-carboxylic acids and aromatic compounds. It is a waste product available in large amounts (80 000 t/year from cork production alone)	None
Pullulan	α-$(1 \rightarrow 4)$-linked glucose trimer, linked by α-$(1 \rightarrow 6)$ bonds, fungal polymer that could be produced in plants	Edible films, fibers
Hyaluronic acid	Repeating disaccharide unit consisting of an *N*-acetyl-hexosamine and a hexose or hexuronic acid, either or both of which may be sulfated	Surgery

netic lesion on starch biosynthesis are much more complex than expected. In addition, the structures of starch and starch granules are still not completely understood [13].

Most of the research on the in planta modification of starch was carried out with food applications in mind, but could aid the production of thermoplastic starch (TPS). For example, high-amylose TPS was reported to have better properties than "standard" TPS: films were less sensitive to water, and less subject to cracking and shrinking [14]. One of the main barriers to appli-

Table 2 Thermoplastic starch

Chemical composition	Amylose: linear α-(1,4)-linked D-glucose polymer, molecular weight 10^5–10^6 Amylopectin: α-(1,4)-linked D-glucose polymer, α-(1,6)-branches, molecular weight 10^7–10^9
Annual production	40 000 t/year
Price	0.20–0.50 €/kg, depending on source
Main sources	Maize, potato, wheat, cassava, rice, pea, waxy and amylo maize, etc.
Main industrial uses	Diapers, cardboard, paper, fabrics, plastics, plaster, water treatment, detergent, oil drilling, filler for tires
Main producers of TPS or starch foam	Novamont, BIOP, Biotec, Rodenburg Biopolymers, Green Light Products, National Starch and Chem., Earthshell
Main use as bioplastic	Foams (for the loose fill foam market), mulch films, shopping bags, moldable products (pots, cutlery, fast food packaging)
Advantages	Cheap, widely available, many variant starches, many functional groups for derivatization, grafting, and interaction with plasticizers
Disadvantages	Mechanically weak, brittle, moisture sensitive, complex heterogeneous multiphase materials, sensitive to retrogradation, poor interaction with plasticizers and hydrophobic polymers, suitable only for short life applications (20% of the market), slow production rates in plastic film equipment
Important issues	1. The potential for starch bioplastics is several million t/year 2. Starch is a complex material (granule structure, amylose vs. amylopectin, crystallinity, chain-length) that is still not fully understood 3. Almost everything has been tried to improve properties 4. In planta modification of starch involves transgenic food plants 5. Starch yield should not be affected by modifications

cations of TPS is its high moisture sensitivity and difficult processing. Such problems can be remedied by chemical derivatization, e.g., by introducing ester and ether-groups [15]. Blending TPS with polycaprolactone or other biodegradable hydrophobic polymers, or by coating TPS films with a water-barrier [16] is also extensively used. Both types of research have been carried out for many decades and are covered by numerous patents. Is it possible and worthwhile to aim for chemical derivatization or blending in planta? Theoretically, a linear starch chain can be decorated with side-groups using enzymes that are co-expressed with the starch synthesizing enzymes. For example, enzymes could be used that O-acetylate cell wall polysaccharides [17]. It must be noted that such modifications would have to take place before the polysaccharide becomes part of the starch granule, because after granule formation, the amylose and amylopectin chains are probably not accessible for modification.

Other factors to keep in mind are the potential deleterious consequences of starch modifications. Starch granule structure is likely to be affected by the modifications, which in turn will affect the amount of the starch produced by the plant. Seed germination might be affected, as one of the biological reasons given for O-acetylation is inhibition of cell wall degradation [17]. In addition, the processing steps required to isolate the material could become less efficient. Here, it is useful to consider the effects of the currently available in planta modifications. For example, reducing the number of branching points in amylopectin resulted in potato plants producing smaller but more numerous tubers [11]. On the other hand, waxy and high-amylose maize variants are commercially grown [12]. In considering modified-starch production, it should be noted that all starch plants are food plants, and that many of the modifications discussed entail the use of transgenic plants, which may not be well accepted by the public when crops that could enter the food chain are used.

3.2 Rubber

3.2.1 Rubber from *Hevea brasiliensis*

Natural rubber (hereafter simply referred to as rubber in contrast to synthetic rubber) consists mainly of *cis*-polyisoprene, with many minor additional components that are the key to the superior properties of this material compared to all synthetic rubbers. Nearly 80% of all rubber is produced by only three countries (Malaysia, Indonesia, and Thailand), and from one biological source: the Brazilian rubber tree (*Hevea brasiliensis*) (Table 3). The yield per hectare varies from 500 kg/year in smallholder plots to more than 1500 kg/year in large plantations [18].

Rubber is a highly valuable biomaterial: in contrast to the other biopolymers discussed, it is essential for many industrial applications, and cannot be replaced by synthetic materials. For example, heavy-duty tires for trucks, buses, and airplanes, as well as latex products for the medical profession, cannot be made with synthetic rubber. The rapid economic development in Asia, especially in China (the world's largest rubber consumer imported 1.5×10^6 t in 2005) and India, is resulting in strongly rising prices. According to the International Rubber Study Group, the production deficit for 2006 is estimated to be 250 000 t.

Because rubber is essential and one region dominates production, rubber is considered a strategic commodity. In 1934, South American leaf blight (SALB) wiped out the production of rubber in Brazil, and it has not been possible to restart large-scale production due to the endemic leaf blight pathogen *Microcyclus ulei* (the present production on marginal lands in Brazil, where

Table 3 Natural rubber

Chemical composition	Major component: *cis*-1,4-polyisoprene, Minor components: proteins, polysaccharides, minerals
Annual production	9×10^6 t/year
Price	Up to 1.8 €/kg, depending on grade
Main source	*H. brasiliensis* (rubber tree)
Producing countries	Main: Indonesia, Malaysia, Thailand Minor: Sri Lanka, India, China, Ethiopia, Nigeria, Brazil
Main uses	Tires, gloves, thread, condoms
Advantages	High resilience, long fatigue life, very good tensile and tear properties, good creep and stress relaxation resistance, efficient heat dispersion, low-temperature flexibility, good balance of properties for demanding mechanical applications
Disadvantages	Compared to some expensive synthetic rubbers: doesn't age well, inferior resistance to sunlight, oxygen, ozone, solvents and oils, variable quality due to local production, re-use is difficult
Alternative plant sources	Guayule (10 000 t/year in 1910, efforts during WWII and oil-crisis), Russian dandelion (WWII efforts), Goldenrod (R&D in 1930s)
Related natural materials	Gutta percha and Balata (poly-*trans*-isoprene) Chicle (mixture of *cis* and *trans*)
Synthetic alternatives	Synthetic rubber (total 10.4×10^6 t/year): styrene–butadiene copolymers (2.4×10^6 t/year), acrylonitrile-butadiene copolymers, and others
Important issues	1. *H. brasiliensis* is a genetically extremely narrow crop: SALB could destroy rubber production in South-East Asia 2. Rubber price strongly increases, a 25% shortfall in production is expected in 15 years 3. Increased competition for land-use by palm-oil plantations (for biodiesel and food applications) 4. Rubber production from *H. brasiliensis* cannot be mechanized, and work-force is getting more expensive 4. Synthetic rubber alternatives are non-renewable 5. Allergenic hypersensitivity to *H. brasiliensis* rubber is increasing

SALB is less of a problem, is only 96 000 t/year). SALB could cause a disaster in Asia, as *H. brasiliensis* is genetically very homogeneous: the millions of hectares of rubber plantations are all derived from a small sample of seeds collected in Brazil by Dr. Henry Wickham in 1876 [19]. Production in Africa is quite limited, although climate and soils would permit large-scale production.

Attempts in the 1980s and 1990s in Brazil to develop SALB-resistant *Hevea* clones did not meet with success. Although some progress has been made [20], all promising lines finally succumbed to the fungus in the field [21]. Apart from efforts in Asia on common plant diseases, yield, and agronomics, *H. brasiliensis* is studied in France and Brazil to generate leaf-blight resistant varieties, increased yield, and altered properties. Recently, efficient transformation of calli and regeneration of plants was shown to be possible [22]. However, the narrow genetic base, prolonged breeding cycles and juvenile period, and highly heterozygous nature of *H. brasiliensis* make breeding complex, time-consuming and labor-intensive. In view of the critical importance of rubber, these efforts appear extremely limited: it makes sense to investigate alternative production methods.

3.2.2 Guayule as an Alternative Source of Rubber

Only one other plant has been used in large-scale commercial production of rubber. In 1910, 10 000 t/year of rubber was produced from natural stands of the guayule shrub (*Parthenium argentatum*) (Table 4) [23]. As production from *H. brasiliensis* became more efficient, and natural stands of guayule were exhausted, this production strategy was gradually abandoned. Guayule was studied intermittently for strategic reasons during WWII and the oil-crisis, and more recently also because many consumers are allergic to *H. brasiliensis* rubber, but not to guayule rubber [24]. Over the years, guayule breeding efforts have improved rubber yield to 1000 kg/ha/year (compared to 1500 kg/ha/year for *H. brasiliensis*) [23]. In Europe, guayule has not attracted much attention, except for limited cultivation studies in Spain and Greece. As the plant is quite vulnerable to cold winters, the initial priorities might include the development of more hardy strains that can be grown in Southern Europe, or the identification of more suitable regions for growing this crop (e.g., North Africa). General research areas requiring attention are breeding for higher yield, harvesting methods, processing, and co-product utilization [25].

3.2.3 Natural Rubber from Plants Growing in Temperate Climates

The last major research activity of Thomas Edison was the development of natural rubber production from Goldenrod (*Solidago virgaurea minuta*) (Table 4). Extensive research proved that Goldenrod, a common weed growing to an average height of 1 m, produced 5% yield of latex. Through hybridization, Edison produced Goldenrod in excess of 3 m, yielding 12% latex. However, Goldenrod rubber never went beyond the experimental stage (http://en.wikipedia.org/wiki/Goldenrod), mainly because the rubber was of low quality.

Another potential source of rubber is the Russian dandelion (*Taraxacum kok-saghyz*). The root is a source of high quality latex (used for making rubber during WWII) with yields of between 150 and 500 kg/ha, and 45 kg of rubber per ton of roots [26, 27]. Unlike guayule latex, dandelion latex is prob-

Table 4 Alternative sources of poly-*cis*-isoprenes

Source	Production in t/year (year)	Price €/kg	Current R&D related to rubber	Refs.
Rubber tree *H. brasiliensis*	9×10^6 (2005)	1.80	Resistance to SALB, rubber polymerase	[22]
Guayule shrub *P. argentatum* Gray	10 000 (1910)	n.a.	Processing technology, rubber polymerase	[30]
Goldenrod *S. virgaurea minuta*	Demonstration project (1931)	n.a.	None	[26]
Russian dandelion *T. kok-saghyz*	WWII emergency projects USSR/USA, 3000 (1943)	n.a.	Domestication	[27]

ably less suitable for medical applications as it contains many proteins that are apparently related to *H. brasiliensis* latex proteins [28]. However, it has a shorter life-cycle than guayule, over 50 000 EST-sequences are available, and it has a relatively small genome [29]. Research carried out in the 1930s and 1940s indicate that although high quality rubber could be produced, the agronomics are not favorable [27].

Production of rubber in transgenic sunflower, lettuce, or chicory has been considered [28]. However, the use of a transgenic food-crop for the production of rubber may not be acceptable in the EU.

4 Protein-Based Bioplastics or Biopolymers

Three groups of protein-based plastics and biomaterials can be distinguished: (1) derivates of natural plant proteins obtained as co-products of starch, vegetable oil, or biofuel production; (2) fibrous proteins with potential uses in engineering (e.g., spider silk, mussel adhesive protein, collagen, elastin); and (3) non-ribosomally produced polypeptides (e.g., cyanophycin and polylysine).

4.1 Protein Co-products

Examples of materials that can be derived from natural plant proteins are plastics and resins based on zein (corn protein) [31], soy protein [32], and gluten from wheat [33]. These materials are typically produced by cross-linking proteins with glutaraldehyde, formaldehyde, or other chemicals, in

Table 5 Protein co-products potentially available for the production of bioplastics or biopolymers

Protein	Total crop harvest (t/year)	Protein content %	Uses of protein as material	Price and volume
Zein (maize)	692×10^6	4	Films, bioplastic, fibers	10–20 €/kg, < 1000 t/year
Soy protein (soybean)	209.5×10^6	38–45	Films, extruded foams, injection molded products	Price is slightly higher than conventional plastics [34]
Gluten (wheat)	626×10^6	9–15	Films, coatings, bioplastics, resins	Not available
Switchgrass leaf protein	Not available	10	Not available	Not available

combination with starch, polyphosphate, or other fillers. Zein is the major protein in corn. In 1950, about 2700 t/year of zein plastics (glossy, scuff-proof, grease-proof coatings) and 2200 t/year of Vicara fiber were produced. If produced on the same scale as in the 1950s, and as a by-product from ethanol production, zein would cost about 2.5 €/kg [31], the actual cost now being ten times higher. Henry T. Ford used soy protein as a source of bioplastics to construct car parts. However, after a brief bloom in the 1930s and 1940s, petroleum-based plastics replaced protein-based plastics, in part because of microbial degradation and water permeability issues [32]. Gluten-based bioplastics suffer from the same general problems and are also currently too expensive for large-scale use.

The amount of protein co-products from future large-scale biofuel production (potentially millions of tons of protein per year from Switchgrass or *Miscanthus*) can be expected to greatly exceed the amount that can be absorbed by the food and feed markets, enabling the development of a protein-based bioplastics industry (Table 5). The role for genetic engineering specifically to improve bioplastics derived from these proteins appears quite limited, especially if the primary goal is biomass production.

4.2 Fibrous Proteins

The second group of protein biopolymers consists of fibrous proteins that are typically composed of short blocks of repeated amino acids. Silk, elastin, adhesin, and numerous other fibrous proteins show great promise in that these materials have unique strength-to-weight, elastic, or adhesive properties [35, 36]. These are potentially very attractive materials, but expensive and labor-intensive to produce from their natural sources. Therefore, quite some effort has gone into the heterologous production of these proteins. In most cases, microorganisms (but also cell cultures, animals, and plants) were tested as host. Problems such as clone instability because of repetitive sequences, inclusion bodies, and difficult processing have thus far prevented breakthroughs. It has proven difficult to obtain materials (fibers, glues, elastic tissue) from recombinant material with the same quality as the original material except on a small scale. It must be kept in mind that the properties of fibers such as silk depend in large part on how different types of proteins are assembled and spun together. Thus, beyond the production of the individual protein components, advances in microspinning technologies are essential. At present, production of fibrous proteins in plants suffers from the same problems, i.e., low yield [37] and difficult processing (e.g., spinning of heterologously produced silk). Concerning yield, approaches such as seed-specific expression and the use of ER-targeting sequences may provide valuable solutions [38].

Genetic engineering can also be used to produce completely new materials such as block-copolymers, combinations of different proteins like silk and elastin, completely synthetic sequences with even better properties, and thus perhaps also sequences optimized for production in specific organisms, including plants. Heterologous expression in plants would enable production on a much larger scale and open up new markets. However, the question should be asked if any of the fibrous proteins has a (potential) market size that would justify the development of a transgenic germplasm. Moreover, it should be noted that one of the most interesting aspects of the fibrous proteins is the ability to specify properties through the DNA template. This allows tailoring to specific applications and processing, but at the same time clearly favors production in more flexible organisms, such as bacteria or yeast. In addition, much higher product concentrations can be attained in these organisms without compromising growth, the downstream processing is likely to be easier, as is the genetic engineering (especially in view of the multitude of different proteins in this class). In addition, if these proteins are to be used as high-end engineering materials, and thus needed on a relatively small scale, the fermentation costs are less relevant than the material properties. Thus, production in plants should be envisaged only if the fibrous protein is to be used on a commodity-scale.

4.3 Non-ribosomal Polypeptides

The third group of protein biopolymers consists of non-ribosomally produced polypeptides such as cyanophycin, a protein-like copolymer composed of a polyaspartate backbone and arginine side-groups produced by cyanobacteria and a few non-photosynthetic bacteria, as well as polylysine and polyglutamate. The latter are now used in food but have many potential applications ranging from hydrogels, biochip coatings, drug carriers, cryoprotectant, etc. [39, 40]. Polyaspartate derived from cyanophycin can be used as superadsorbant or antiscalant (see references cited in [41]). Recombinant *E. coli* can produce cyanophycin up to 29% of the cell dry weight on protamylasse, a waste product of starch production from potato [42]. Transgenic plants have been created that contain up to 1.1% cyanophycin dry weight [41]. Due to the low-price applications of these compounds, the critical question is whether production levels in plants can be high enough for cheap production. Again, production in crop plants versus bacteria makes sense only if the polypeptides or their derivatives are used as commodities.

5
Poly-β-hydroxyalkanoates

Poly-β-hydroxyalkanoates (PHA) are polyesters naturally produced by microorganisms, primarily as carbon and energy storage material. The polymer properties depend strongly on the nature of the monomer, which can range from linear C_4–C_{16} β-hydroxy fatty acids, to β-hydroxy acids substituted with aromatic rings, other functional-groups, or containing double-bonds. The simplest PHA, poly-β-hydroxybutyrate (PHB), is a relatively hard and brittle material with a melting point slightly below the thermal decomposition temperature [43]. Inclusion of C_5-monomers gives slightly better properties. Adding small amounts of longer monomers (C_6 and longer) has resulted in materials with further improved processing and material properties [44]. PHA consisting of higher molecular weight monomers (C_6–C_{16}, referred to as medium-chain PHA or mclPHA) typically are rubber-like materials with an amorphous soft-sticky consistency (Table 6) [45].

PHAs are very attractive polymers for consumer products such as bottles, films, and fibers, due to their water and air impermeability, as a source of chiral monomers, and as components of paints [46]. If it is possible to produce PHAs at a cost of 1–2 €/kg, many of these potential applications become commercially viable. Presently, PHAs produced by microbial fermentation are clearly too expensive, estimated at 10 €/kg [8]. However, according to some industry specialists, the lower price range is feasible with current, large-scale and fully integrated bioreactors, and downstream processing technology. Present efforts to develop cellulosic ethanol [47], and the rapid development of biogas technology to convert waste biomass into heat and electricity, should make PHA fermentation technology much more energy and CO_2 efficient [48].

As a potential large-scale commodity, it is logical to consider production of PHA in plants [49]. The critical question to ask is: can PHA produc-

Table 6 Poly-β-hydroxyalkanoates

Chemical composition	Linear polyesters of 3-hydroxyalkanoates and related hydroxy acids 1. Poly-β-hydroxybutyrate (PHB), high crystallinity 2. PHB-co-valerate (PHBV), high crystallinity 3. PHB-co-hexanoate (PHBH or Nodax), moderate crystallinity 4. Medium-chain PHA (mclPHA), C_6–C_{16} monomers, elastomers, low crystallinity
Annual production	< 1000 t/year (Metabolix, Biomer, Biomatera, Kaneka) Monsanto stopped production of PHBV in 1998 ADM & Metabolix announced construction of a 50 000 t/year plant in 2006
Price	1.5 €/kg (expected), currently 10–20 €/kg
Main source	Bacterial fermentation using sugars and oils as starting material
Main producing countries	USA, Brazil, Germany, Japan, China, Thailand, presently all at a very small scale
Main (industrial) uses	1. Thermoplasts for bottles, packaging material, cutlery, cups, bags, mulching films 2. Latex for coatings and films 3. Blending with other biodegradable polymers 4. For mclPHAs: source of monomers, paints, pressure-sensitive adhesives, biodegradable cheese coatings, and biodegradable rubbers
Advantages	1. Hydrophobic and moisture-resistant compared to other biopolymers 2. Choice of feed strategy and host organism allows many different monomer compositions, resulting in a wide range of properties: for example, PHBH (Nodax) is easier to process than PHBV due to lower melting temperature and lower crystallinity, and has greater toughness and ductility 3. High oxygen impermeability 4. Processing on conventional equipment for polyolefins possible
Disadvantages	General: high production costs, hydrophobicity makes blending with cheap hydrophilic polymers such as starch and proteins difficult PHB: brittle, stiff, decomposes just above melting temperature, unfavorable aging PHBV: slightly lower melting temperature than PHB, long processing times mclPHA: weak, sticky, rubbery
Related materials	Polylactate (PLA), polycaprolactone (PCL), other polyesters produced by condensation of diacids and diols, or hydroxy acids
Important issues for production of PHAs in plants	1. Transgenic food plants will not be accepted in the EU 2. LCA and land-use favor plant GMO over bacterial fermentation 3. Deleterious effects on plant growth at high PHA levels 4. Lack of control over monomer composition 5. Processing 6. Stability in harvested material

tion in plants compete with established and future fermentation methods? If the answer is positive, many technical hurdles need to be addressed. These issues are generally related to production levels in plant tissue, control over monomer composition, deleterious effects of PHA production on plant growth, and processing technology. In *A. thaliana* PHB levels of 40% based on dry weight of leaves were obtained, but plant growth was severely affected [50]. It is conceivable that better control over targeting (in the cell organelle or plant part where the PHA is produced) will solve this problem. PHB up to 8% in seeds of rape has been reported without deleterious effects on seed germination or viability [51]. Production of PHA containing several types of monomers (from C4 to C16) has been reported for a variety of plants [51].

Polyesters with different monomer compositions are easily obtained by using different bacterial hosts, feeding regimes, and co-feeding specific monomers [45]. The breadth and precision in monomer composition that can be attained by bacterial fermentation appears to be hard to replicate in plants, because in some cases at least two independent metabolic pathways supplying the intermediates would have to be quantitatively controlled during production of the polymer if it is to contain two or more monomers in defined ratios. This requires a much better understanding of and control over metabolic pathways and fluxes in plants than is presently available.

Isolation of PHA from plant tissues is bound to be more complicated than isolation from bacteria where there is no need to break up tissue. Further, much higher concentrations can be reached in bacteria without affecting the viability of the host organisms (up to 85% for PHB). The timing of production in a bioreactor is also much easier as the typical substrates for bacterial growth (sugars and oil-containing wastes, or purified compounds) can be stored, and production can take place throughout the year. A related issue is whether the PHA is stable in plant material after harvesting: if not, the plants must be processed immediately after harvest (this is also an issue for rubber and heterologously produced proteins).

Taking into account the cost to create a transgenic plant and the time required to generate a commercial germplasm, it seems advisable to concentrate on the production of only one or two standard PHA polymers in plants (PHB and perhaps mclPHA), leaving production of the wide range of other PHA polymers to fermentation schemes.

Which plants are most suitable for PHA production? Since production of PHAs in plants by definition involves the use of transgenic plants, from the EU perspective it is best to focus on a non-food crop, such as Switchgrass chosen by Metabolix (www.metabolix.com), energy-crops such as *Miscanthus*, or a non-food oil crop such as *Crambe*.

6 Conclusions

Bioplastics and materials derived from biopolymers typically have low value applications, but potentially large markets (the exceptions are fibrous proteins, which are more valuable but have smaller markets). In most cases, the bioplastics have to compete with petrochemical plastics on price and properties.

The role for transgenic plants differs strongly depending on the biopolymer. Starch has rather unfavorable properties for use as a thermoplastic except for its low price. It is difficult to envisage in planta modifications that will drastically improve the material properties of starch. In the case of natural rubber, the first priority appears to be the development of alternative crops such as guayule or Russian dandelion. Genetic engineering may help the development of improved germplasm. The only polypeptide where transgenic plants appear useful are the non-ribosomal protein cyanophycin; it seems unlikely that protein co-products of biofuels or food production can be modified to improve their usefulness as a bioplastic (although other applications can be considered). For fibrous proteins, it makes more sense to concentrate on production in microorganisms, because markets are small and many different proteins (and derived sequences) must be considered. Finally, for PHA it makes sense to concentrate on one or two specific PHAs, for example PHB and mclPHA.

Acknowledgements This work was funded by a grant from the European Union 6th Framework Programme (EPOBIO; SSPE-022681).

References

1. Gaskell G, Allansdottir A, Allum N, Corchero C, Fischler C, Hampel J, Jackson J, Kronberger N, Mejlgaard N, Revuelta G, Schreiner C, Stares S, Torgersen H, Wagner W (2006) Europeans and biotechnology in 2005: patterns and trends, Eurobarometer 64.3. European Commission's Directorate-General for Research, Brussels, Belgium
2. Mecking S (2004) Angew Chem 43:1078
3. Thakor N, Luetke-Eversloh T, Steinbuechel A (2005) Appl Environ Microbiol 71:835
4. Kurdikar D, Fournet L, Slater SC, Paster M, Gruys KJ, Gerngross TU, Coulon R (2001) J Ind Ecol 4:107
5. Dornburg V, Lewandowski I, Patel M (2004) J Ind Ecol 7:93
6. Luinstra G, Almendinger M, Rieger B (2005) US Patent 7 019 107
7. The National Non-Food Crops Centre UK (2005) The promotion of non-food crops. Report number IP/B/AGRI/ST/2005-02. Directorate General Internal Policies of the European Union, Brussels, Belgium
8. Crank M, Patel M, Marscheider-Weidemann F, Schleich J, Hüsing B, Angerer G (2004) PRO-BIP. Techno-economic feasibility of large-scale production of bio-based polymers in Europe. European Commission's Institute for Prospective Technological Studies, Utrecht, Karlsruhe

9. Van Camp W (2005) Curr Opin Biotechnol 16:147
10. Blennow A (2003) Recent Dev Carbohydrate Res 1:95
11. Blennow A, Wischmann B, Houborg K, Ahmt T, Jorgensen K, Engelsen SB, Bandsholm O, Poulsen P (2005) Int J Biol Macromol 36:159
12. Jobling S (2004) Curr Opin Plant Biol 7:210
13. Morell MK, Myers AM (2005) Curr Opin Plant Biol 8:204
14. Van Soest JJG, Essers P (1997) J Macromol Sci, Pure Appl Chem A34:1665
15. Kaplan DL (ed) (1998) Biopolymers from renewable sources. Macromolecular systems – materials approach. Springer, Berlin Heidelberg New York
16. Bastioli C (2005) Handbook of biodegradable polymers. Rapra Technology, Shawbury
17. Pauly M, Scheller HV (2000) Planta 210:659
18. Balsiger J, Bahdon J, Whiteman A (2000) The utilization, processing and demand for rubberwood as a source of wood supply. FAO, Forestry Policy and Planning Division, Rome, Italy
19. Davis W (1997) The rubber industry's biological nightmare. Fortune Magazine, 4 August 1997
20. Le Guen V, Lespinasse D, Oliver G, Rodier-Goud M, Pinard F, Seguin M (2003) Theor Appl Genet 108:160
21. Lespinasse D, Grivet L, Troispoux V, Rodier-Goud M, Pinard F, Seguin M (2000) Theor Appl Genet 100:975
22. Blanc G, Baptiste C, Oliver G, Martin F, Montoro P (2006) Plant Cell Rep 24:724
23. Ray DT, Coffelt TA, Dierig DA (2005) Ind Crops Prod 22:15
24. Cornish K (1996) US Patent 5 580 942
25. Nakayama FS (2005) Ind Crops Prod 22:3
26. Polhamus LG (1962) Rubber: botany, production, and utilization. Leonard Hill, London
27. Whaley WG, Bowen JS (1947) Russian dandelion (Kok-Saghyz). An emergency source of natural rubber. United States Department of Agriculture, Washington DC
28. Cornish K, McMahan CM, Pearson CH, Ray DT, Shintani DK (2005) Rubber World 233:40
29. Falque M, Keurentjes J, Bakx-Schotman JMT, van Dijk PJ (1998) Theor Appl Genet 97:283
30. Mooibroek H, Cornish K (2000) Appl Microbiol Biotechnol 53:355
31. Lawton JW (2002) Cereal Chem 79:1
32. Mohanty AK, Liu W, Tummala P, Drzal LT, Misra M, Narayan R (2005) Soy protein-based plastics, blends, and composites. In: Mohanty AK, Misra M, Drzal LT (eds) Natural fibers, biopolymers, and biocomposites. CRC, Boca Raton, FL, p 699
33. Pallos FM, Robertson GH, Pavlath AE, Orts WJ (2006) J Agric Food Chem 54:349
34. Wondu Holdings (2004) Bioplastics supply chains – implications and opportunities for agriculture. RIRDC publication no: 04/044. Australian Government, Rural Industries Research and Development Corporation, Kingston, ACT, Australia
35. Sanford K, Kumar M (2005) Curr Opin Biotechnol 16:416
36. Scheibel T (2005) Curr Opin Biotechnol 16:427
37. Scheller J, Conrad U (2005) Curr Opin Plant Biol 8:188
38. Yang J, Barr LA, Fahnestock SR, Liu Z-B (2005) Transgenic Res 14:313
39. Shih IL, Shen MH, Van YT (2006) Biores Technol 97:1148
40. Shih IL, Van YT (2001) Biores Technol 79:207
41. Neumann K, Stephan DP, Ziegler K, Huhns M, Broer I, Lockau W, Pistorius EK (2005) Plant Biotechnol J 3:249
42. Elbahloul Y, Frey K, Sanders J, Steinbuchel A (2005) Appl Environ Microbiol 71:7759

43. Lenz RW, Marchessault RH (2005) Biomacromolecules 6:1
44. Noda I, Green PR, Satkowski MM, Schechtman LA (2005) Biomacromolecules 6:580
45. Van der Walle GAM, De Koning GJM, Weusthuis RA, Eggink G (2001) Adv Biochem Engin Biotechnol 71:263
46. Witholt B, Kessler B (1999) Curr Opin Biotechnol 10:279
47. Farrell AE, Plevin RJ, Turner BT, Jones AD, O'Hare M, Kammen DM (2006) Science 311:506
48. Kim S, Dale BE (2005) Int J LCA 10:200
49. Moire L, Rezzonico E, Poirier Y (2003) J Plant Physiol 160:831
50. Bohmert K, Balbo I, Kopka J, Mittendorf V, Nawrath C, Poirier Y, Tischendorf G, Trethewey RN, Willmitzer L (2000) Planta 211:841
51. Houmiel KL, Slater S, Broyles D, Casagrande L, Colburn S, Gonzalez K, Mitsky TA, Reiser SE, Shah D, Taylor NB, Tran M, Valentin HE, Gruys KJ (1999) Planta 209:547

Invited by: Professor Sautter

Adv Biochem Engin/Biotechnol (2007) 107: 153–172
DOI 10.1007/10_2007_054

Published online: 31 March 2007

Plastoglobule Lipid Bodies: their Functions in Chloroplasts and their Potential for Applications

Felix Kessler[1] (✉) · Pierre-Alexandre Vidi[2,2]

[1]Institute of Botany, University of Neuchâtel, Emile-Argand 11, CP158, 2009 Neuchâtel, Switzerland
felix.kessler@unine.ch

[2]*Present address:*
Department of Medicinal Chemistry and Molecular Pharmacology, Purdue University, 575 Stadium Mall Dr., West Lafayette, IN 47907, USA

Abstract Plastoglobules are plant lipid bodies localized inside plastids. They have long been considered as mere lipid storage compartments. However, ultrastructural and proteomic data now suggest their involvement in various metabolic pathways, notably the biosynthesis of tocopherols. In this work, the current knowledge on the structure and functions of plastoglobules is reviewed. On the basis of similarities between plastoglobules and seed oleosomes, the potential of plastoglobules for bioengineering applications is discussed.

Keywords Lipid bodies · Plastids · Plastoglobules · Protein purification · Tocopherols

Abbreviations

ABA	Abscissic acid
ADRP	Adipocyte differentiation related protein
AOS	Allene oxide synthase
CCD	Carotenoid cleavage dioxygenase
FBA	Fructose-6-bisphosphate aldolase
JA	Jasmonic acid
NCE	9-*cis*-epoxycarotenoid
OPDA	12-oxo-phytodienoic acid
PAP	Plastid lipid-associated protein
PGL	Plastoglobulin
PQH_2	Plastoquinone
PSII	Photosystem II
ROS	Reactive oxygen species
TAG	Triacylglycerol
VTE	Vitamin E deficient

1 Introduction

Plastids form a group of plant-specific organelles. All plastid types initially derive from proplastids, which are small, undifferentiated organelles abundant in meristematic tissues. As suggested by the etymology of their name ("plassein", the Greek for "to mould" or "shape"), these organelles are highly plastic, both in structure and function. Developmental or environmental stimuli cause plastids to differentiate into specialized plastid types such as photosynthetic chloroplasts, colored chromoplasts or storage plastids (amyloplasts and elaioplasts) [1–3]. Plastids are seen as the result of an ancient endosymbiotic event where a cyanobacterial ancestor invaded a primitive eukaryotic cell [4].

Although plastids are semi-autonomous organelles that retained genetic material as well as transcription and translation machineries, the vast majority of their proteins are encoded in the nucleus and synthesized as cytosolic preproteins with N-terminal transit sequences that need to be imported in plastids (see [5–7] for recent reviews). The transit sequences are recognized by translocon at the outer (Toc) and inner (Tic) chloroplast membranes. The Tic-complex consists of a variety of translocon (Tic20, -21, -22, -40, -110) as well as redox and calcium regulatory components (Tic32, Tic55). Tic22 and Tic110 have been proposed to function as components of the import channel. In addition Tic110, together with Tic40 may function as a co-chaperone recruiting ClpC, cpn60 and, possibly, Hsp70 chaperones to the import site to assist in the folding of the newly imported proteins. The Toc-complex consists of Toc75, a protein conducting channel, and of two homologous GTP-binding proteins, Toc159 and Toc34 functioning as transit sequence receptors exposed at the chloroplast surface. These are encoded by small gene families in Arabidopsis [8]: Toc159 has four homologs in Arabidopsis (AtToc159, -132, -120 and -90) and Toc34 has two

(AtToc34 and -33). Recent genetic and biochemical studies indicate that different combinations of Toc GTPases facilitate the import of specific classes of substrates [6], AtToc159 and AtToc33 together being responsible for the import of constituent proteins of the photosynthetic apparatus.

Although plastids have different shapes and functions, they retain a similar architecture. A double membrane, termed the envelope, delimits the boundary of the organelles. The inside of plastids is composed of an aqueous matrix, the stroma, and in chloroplasts of an extended membrane system, the thylakoids. Lipid bodies, referred to as osmiophilic bodies or plastoglobules, are ubiquitous structures in the stroma. Plastoglobules have been implicated in plant stress response, chloroplast-to-chromoplast transition and thylakoid disassembly in senescing tissues. It is widely accepted that they serve as lipid reservoirs in plastids. Chloroplast plastoglobules have a particular lipid composition and are notably enriched in tocopherols (vitamin E). They are coated with structural proteins from the PAP/PGL/fibrillin family and recent proteomic studies have identified enzymes as genuine plastoglobule components.

Here, we will review the structure, composition and physiological relevance of plastoglobules with a focus on chloroplasts. The potential of plastoglobules for bioengineering applications will then be discussed.

2
Structure and Composition of Plastoglobule Lipid Bodies

Analysis of chloroplast ultrastructure by electron microscopy revealed the presence of lipid bodies ("osmiophilic globuli") in the stroma [9–12]

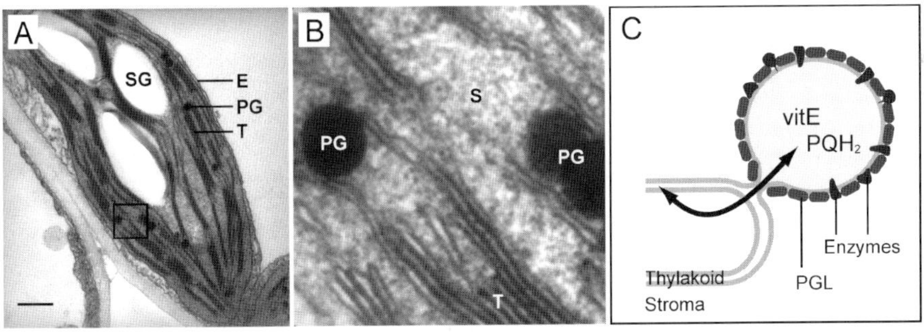

Fig. 1 Structure of plastoglobules. **A** Transmission electron micrograph showing a chloroplast in an Arabidopsis leaf. *Scale bar*: 500 nm. **B** Enlargement from **A**. PG, plastoglobule; SG, starch granule; E, chloroplast envelope membranes; T, thylakoids; S, stroma. **C** Structural model showing the association of a plastoglobule with thylakoid membranes. The half-lipid bilayer surrounding the hydrophobic core of plastoglobule (*light gray*) is continuous with the stroma-side leaflet of thylakoids. Dynamic exchanges of lipids between the compartments are proposed. Structural plastoglobulin proteins (PGL, *red*) as well as enzymes from various metabolic pathways (*blue*) are associated at the periphery of plastoglobules. VitE, vitamin E; PQH_2, plastoquinone

(Fig. 1A,B). The diameter of these bodies, hereafter termed plastoglobules, ranges from 30 nm to 5 μm and varies in different species, plastid types, developmental stages and physiological conditions [13]. In certain species, very large plastoglobules accumulate (e.g. [12, 14]), whereas in others plastoglobule clusters are formed (e.g. [15] as well as personal observations; see Fig. 2A). Although plastoglobule size and abundance are highly variable, the lipid bodies are ubiquitously found in plastids of land plant species, alga [16] and in cyanobacteria [17]. Plastoglobules were proposed to occur in the stroma without connections to other plastid membrane systems [13]. However, only a small portion of chloroplast plastoglobules were liberated from

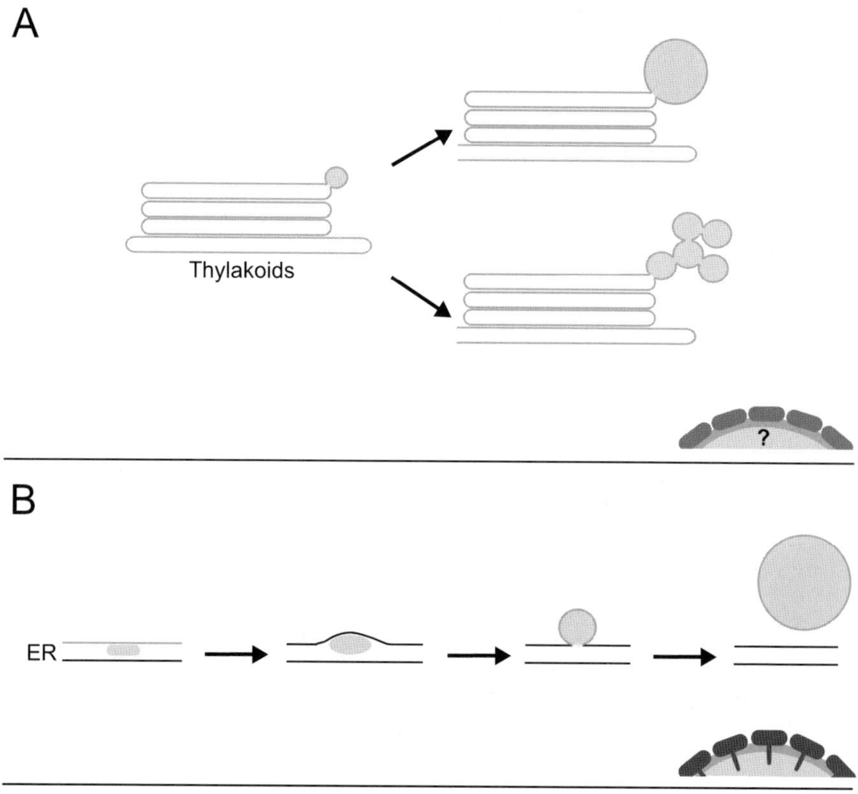

Fig. 2 Models for the formation of plastoglobules (**A**) and oleosomes (**B**). **A** Formation of plastoglobules at the thylakoid membranes occurs through a blistering process. Various developmental and environmental cues including photooxidative stress, senescence and chromoplast differentiation induce enlargement of plastoglobules or the formation of plastoglobule clusters. **B** Oleosome biogenesis in the cytosol. Accumulation of TAGs in sub-domains of the ER precedes budding of oil bodies (after [84]). Oleosins (*purple*) are anchored in oleosomes by a hydrophobic stretch in the manner of a drawing pin, whereas plastoglobulins (*red*) may associate with polar lipids at the periphery of plastoglobules

thylakoid membranes in fractionation experiments [18]. Moreover, recent electron tomographic reconstruction of chloroplasts [15] clearly showed association of plastoglobules with thylakoids, the half-lipid bilayer surrounding the globule forming a continuum with the stroma-side leaflet of thylakoid membranes (see Fig. 1C).

2.1 Lipid Composition of Plastoglobules

The main components of chloroplast plastoglobules are triacylglycerols (TAG) and prenylquinones. Plastoquinone (PQH_2) and tocopherols (vitamin E) are the major prenylquinone constituents while phylloquinone (vitamin K) is present in slight amounts [9, 10, 19, 20, 22]. Traces of chlorophylls and carotenoids (β-carotene, lutein) have also been detected in plastoglobule fractions [9, 10] but have been considered as thylakoid contamination [21]. Similarly, glyco- and phospholipids have been identified in plastoglobule preparations but their genuine association with plastoglobules has been questioned [21]. The lipid composition of plastoglobules from non-green plastids is markedly different from that of chloroplast plastoglobules. Plastoglobules from chloroplasts in senescing leaves, for example, accumulate carotenoid esters, oxidized prenylquinones and free fatty acids [22]. Carotenoid esters are also the major constituents of chromoplast plastoglobules and fibrils [23]. Variations in lipid composition suggest that plastoglobules are highly dynamic structures and that their functions evolve during plastid differentiation.

2.2 Plastoglobulins: Structural Proteins Associated with Plastoglobules

Identification of Plastoglobulins

Early biochemical studies identified nitrogen in purified chloroplast lipid bodies, suggesting the presence of associated proteins [19].

A major peptide of 35 kDa was identified independently by tree groups in bell pepper fruits and designated chromoplast protein B (ChrB, [24]), fibrillin [23] or plastid lipid-associated protein (PAP, [25]). A homologous protein (ChrC) was subsequently identified in cucumber flowers [26]. Expression of fibrillin and ChrC was first proposed to be restricted to chromoplast-containing tissues such as fruits and corollas [23, 27]. However, Pozueta-Romero and colleagues [25] detected fibrillin in leaves and expression of a fibrillin homolog in citrus leaves was reported [28]. Moreover, homologous proteins were identified in leaves from pea (PG1; [29]) and turnip (PAP1-3; [30]), as well as in anthers from rapeseed (BCP32; [31]), indicating that PAP/PGL/fibrillin proteins are not chromoplast-specific and associate with various types of plastid lipid bodies. Several cyanobacterial proteins contain a PAP/PGL/fibrillin motif, indicating an ancient origin of the protein family. Related proteins are, however, absent from bacterial, animal and fungal genomes. Since the abbreviations "PAP" and "FIB" already stand for phosphatidic acid phosphatase and fibrillarin, respectively, we proposed to use the term "plastoglobulin" (PGL) to designate proteins from the PAP/PGL/fibrillin family [32].

Localization and Functions of Plastoglobulins

Association of PGLs with plastoglobules has been shown by immunolabeling on chloroplast ultrathin sections [15, 23, 25, 29, 32] and by tagging the proteins with the green fluorescent protein [32]. Several PGLs were also proposed to associate directly with thylakoid membranes [14, 33, 34]. PGLs do not share sequence homology with known enzymes. Moreover, Deruère and collaborators [23] could reconstitute fibrils in vitro by adding purified fibrillin to chromoplast lipids. Overexpressing fibrillin in tobacco lead to an increase in plastoglobule number and to the formation of plastoglobule clusters [14]. The role of PGLs is therefore probably mainly structural, maintaining the shape of the lipid bodies and preventing their coalescence.

Plastoglobulin Gene Families

In different plant species, PGLs form gene families [30, 35]. The Arabidopsis and rice genomes contains 13 and 8 proteins with a PAP/PGL/fibrillin (PF04755) Pfam profile, respectively. Moreover, recent proteomic studies by Ytterberg et al. [36] and Vidi et al. [32] have shown that at least eight different PAP/PGL/fibrillins are associated with Arabidopsis plastoglobules. As revealed by gene chip expression data, Arabidopsis PGLs have distinct expression patterns [37]. It remains to be determined whether the different PGL isoforms have distinct localizations and/or functions.

2.3
Plastoglobules Contain an Assortment of Enzymes

Although PGLs are the most abundant peptides in fibrils and plastoglobules, SDS-PAGE analysis indicated that at least a dozen different proteins associate with plastid lipid bodies [29, 38]. Analysis of the proteome of Arabidopsis plastoglobules [32, 36] revealed enzymes belonging to various biochemical pathways as genuine plastoglobule components (Table 1). Fructose-bis-phosphate aldolase (FBA) isoforms, an epoxycarotenoid dioxygenase (CCD4), the allene oxide synthase (AOS), the tocopherol cyclase (VTE1), putative lipid-modifying enzymes, as well as members of the ABC1/UbiB family were notably identified in both stud-

Table 1 Proteins with structural properties or (predicted) enzymatic activities associated with plastoglobules

AGI code	Gene name or annotation	Functional category[a]	Refs.
At3g26070	Plastoglobulin (AtPGL25/FIB3a)	Structural protein	[27]
At2g42130	Plastoglobulin (AtPGL30/FIB7b)	Structural protein	[27, 31]
At3g23400	Plastoglobulin (AtPGL30.4/FIB4)	Structural protein	[27, 31]
At2g46910	Plastoglobulin (AtPGL31/FIB8)	Structural protein	[31]
At4g22240	Plastoglobulin (AtPGL33/FIB1b)	Structural protein	[27, 31]
At3g58010	Plastoglobulin (AtPGL34/FIB7a)	Structural protein	[27, 31]
At4g04020	Plastoglobulin (AtPGL35/FIB1a)	Structural protein	[27, 31]
At2g35490	Plastoglobulin (AtPGL40/FIB2)	Structural protein	[27, 31]
At2g21330	F6-BiP aldolase (FBA1)	Sugar metabolism	[27, 31]
At4g38970	F6-BiP aldolase (FBA2)	Sugar metabolism	[27, 31]
At2g01140	putative F6-BiP aldolase (FBA3)	Sugar metabolism	[27, 31]
At2g39730	Rubisco activase	Sugar metabolism	[31]
AtCg00490	Rubisco large subunit (RBCL)	Sugar metabolism	[31]
At5g42650	allene oxide synthase (AOS)	Jasmonic acid biosynthesis	[27, 31]
At4g19170	9-*cis*-epoxycarotenoid dioxygenase (CCD4)	Neoxanthin cleavage reaction	[27, 31]
At4g32770	VITAMIN E DEFFICIENT 1 (VTE1)	Tocopherol biosynthesis	[27, 31]
At5g08740	NADH dehydrogenase-like protein, glutathione reductase, Dihydrolipoamide dehydrogenase, FAD-dependent pyridine nucleotide-disulphide	Unknown	[27]
At5g05200[M]	M ABC1 family	*Quinone synthesis*	[27, 31]
At1g79600	ABC1 family	*Quinone synthesis*	[27, 31]
At4g31390	ABC1 family	*Quinone synthesis*	[31]
At1g71810	ABC1 family	*Quinone synthesis*	[31]
At1g54570	esterase/lipase/thioesterase	*Lipid metabolism*	[27, 31]
At3g26840	esterase/lipase/thioesterase	*Lipid metabolism*	[27, 31]
At1g78140[M]	M UbiE methyltransferase-related	*Quinone synthesis*	[27, 31]
At2g41040	UbiE methyltransferase-related	*Quinone synthesis*	[27, 31]
At3g26060	peroxiredoxin Q	Oxidative stress response	[27]
At1g32220	3-β-hydroxysteroid dehydrogenase/ isomerase	Unknown	[27, 31]
At2g34460	flavine reductase, steroid biosynthesis	Unknown	[27, 31]
At3g10130	SOUL heme-binding family protein	Unknown	[27, 31]
At1g06690	aldo-keto reductase, ANC transporters family signature	Unknown	[27]
At5g08740	pyridine nucleotide-disulfite oxidoreductase (DhnA-like)	Unknown	[31]

Table 1 (continued)

AGI code	Gene name or annotation	Functional category[a]	Refs.
At1g09340[O]	NAD-dependent epimerase/ dehydratase, putative RNA binding protein	Unknown	[31]
At3g63140	NAD-dependent epimerase/ dehydratase, putative RNA binding protein	Unknown	[31]
At5g01730[O]	–	Unknown	[31]
At4g01150	–	Unknown	[31]
At1g28150	–	Unknown	[31]
At1g26090	–	Unknown	[31]
At4g13200	–	Unknown	[27, 31]
At1g52590	–	Unknown	[27]

[a] Italics indicate predicted functions. Sub-cellular localization: AGI codes of proteins for which subcellular localization prediction (TargetP, [103]) was not plastids are labeled with 'M' (mitochondrial prediction) or 'O' (other). AGI code Gene name or annotation Functional category 1

ies. Plastoglobules isolated from red pepper also contained enzymes, namely zeta-carotene desaturase (ZDS), lycopene β-cyclase (LCY-β or CYC-β) and two β-carotene β-hydroxylases (CrtR-β) [36] catalyzing serial reactions in bicyclic carotenoid biosynthesis [39]. These results indicate that plastoglobules in chromoplasts participate in the synthesis of the carotenoids they subsequently sequestrate. The finding indicates that, similarly to cytosolic lipid bodies in yeast or animal cells [40] (see Sect. 4.3), plastoglobules are metabolically active and not mere lipid storage compartments as previously assumed.

3 Proposed Functions of Plastoglobules

3.1 Plastoglobules as Lipid Reservoirs for Thylakoids

The association of plastoglobules with thylakoid membranes [15] suggests that they play a role in thylakoid membrane function, possibly as a reservoir for certain lipids [13]. Indeed, plastoglobules enlarge during thylakoid disassembly in senescing chloroplasts [13, 41–43]. Their accumulation in senescing rosette leaves of Arabidopsis correlates temporally with the activation of diacylglycerol acyltransferase1 (DGAT1) and with enhanced synthesis of

TAGs [44]. Dismantling of thylakoid membranes during senescence allows remobilization of energy for seed production. TAG accumulating in plastoglobules must therefore leave plastids in order to be converted into sugars through β-oxidation and the glyoxylate cycle. Evidence gained from ultrastructural analysis of senescing plastids indicated that plastoglobules are released from gerontoplasts through a blebbing process [42, 45].

3.2 Deposition of Pigments in Chromoplast Plastoglobules

Plastoglobules have also been implicated in chloroplast-to-chromoplast transition. During chromoplast differentiation, plastoglobules enlarge [13] and accumulate esterified carotenoids [21, 23], conferring to fruits and flowers their attractive colors. In certain species, plastoglobules elongate to form fibrillar structures [2, 46].

3.3 Function of Plastoglobules in Plant Stress Response

Morphological Evidence

Enlarged plastoglobules have been described in chloroplasts under conditions resulting in oxidative stress such as drought [14, 33], hypersalinity [47], nitrogen starvation [48], and growth in the presence of heavy metals [49, 50]. In aloe plants exposed to strong sunlight and drought stress, accumulation in leaves of the red carotenoid rhodoxanthin paralleled transformation of chloroplasts into plastoglobule-rich chromoplasts [51]. Studies on spruce and aspen trees have also identified swelling of plastoglobules as part of the physiological response to elevated ozone concentrations [52, 53]. Ageing has also been shown to affect plastoglobule morphology. Older broad bean leaves had significantly larger plastoglobules than younger ones [9]. The same observation was made in rhododendron leaves [54]. Since levels of reactive oxygen species (ROS) are known to rise with time in plastids [55], swelling of plastoglobules in older chloroplasts may again represent a response to the increase of ROS concentration.

Accumulation of Plastoglobulins under Stress Conditions

Up-regulation of several PGLs has been observed as a consequence of various treatments generating photooxidative stress in chloroplasts. These include drought [14, 56–62], cold [60, 62], salt [60, 62], wounding [56, 60], ageing [59] treatment with methyl viologen, [56, 59] and high light stress [14, 57, 59, 60, 63]. Supporting a role of PGLs in stress response, deregulation of these proteins was shown to affect plant growth and stress tolerance.

Overexpression of pepper fibrillin in tobacco enhanced growth under high light intensities as well as drought tolerance [14]. In contrast, antisense potato plants with reduced levels of the PGL C40.4 displayed stunted growth and reduced tuber yield [34]. Recently, Yang et al. [63] showed a correlation between maximal photochemical efficiency of photosystem II (Fv/Fm) and levels of an Arabidopsis PGL in photooxidative stress conditions. Reduction of Fv/Fm values is regarded as photoinhibition of PSII [64]. These results therefore strengthen the view that PGLs directly or indirectly protect the photosynthetic apparatus.

Tocopherol Antioxidants in Plastoglobules

Tocopherols are known to protect membrane lipids from oxidative damage by scavenging radicals and by quenching ROS (reviewed in [65] and [66]). Tocopherols were recently shown to prevent photoinactivation of the PSII [67]. They also protect seed storage lipids from oxidation [68]. Tocopherol synthesis was upregulated after exposure to high light intensities [69, 70] and messenger levels of the tocopherol cyclase (*VTE1*), notably, strongly increased after exposure to strong light [70]. In chloroplasts and cyanobacteria, absence of tocopherols was accompanied by an increased photoinhibition under conditions of photooxidative stress [71, 72]. When reduction of tocopherols and glutathione [70] or zeaxanthin [67] contents were combined, stronger photoinhibition was observed, indicating that photoprotection is guaranteed by a network of antioxidants, including tocopherols. Tocopherols have been detected in all chloroplast membranes and notably in plastoglobules [10, 73]. In these studies, a strong enrichment in prenylquinones was observed in plastoglobules compared to total chloroplast extracts. Moreover, comparison of various plastid types revealed a positive correlation between plastoglobule abundance and prenylquinone contents. Tocopherol measurements in Arabidopsis chloroplast membrane fractions showed that around 50% of the tocopherol pool is localized in plastoglobules, representing a 25-fold enrichment with regard to thylakoids [32].

A Model for the Involvement of Plastoglobules in the Protection of Thylakoids

Changes in plastoglobule morphology and PAP/PGL/fibrillin abundance probably reflect increased needs for antioxidants under stress conditions: Swelling of plastoglobules could indeed allow accumulation of tocopherols (and possibly other antioxidants such as zeaxanthin) and enzymes such as the tocopherol cyclase. A site of action of tocopherols is the thylakoid membrane system. Providing a mechanistic explanation for metabolite trafficking, plastoglobules and thylakoid membranes have been shown to form a continuum [15] potentially allowing tocopherols to diffuse between these two compartments (Fig. 1C). Phylloquinone and PQH_2, which were also measured in plastoglobules, are both involved in the electron transport chain in thylakoid membranes. Like tocopherols, other prenyl quinones probably extensively diffuse in the thylakoid-plastoglobule lipid continuum.

Under photooxidative stress conditions, charge separation at the PSII occurs faster than oxidation of PSII acceptors [74], which leads to photoinhibition. Plastoglobules may therefore represent a compartment to sequestrate reduced quinone pools. Furthermore, the presence of tocopherols in plastoglobules may protect this pool from oxidative damage. Similarly, tocopherols accumulating in chromoplasts plastoglobules and fibrils [75] probably protect carotenoids from oxidation.

Plastoglobules and Signaling Networks

Abscissic acid (ABA) is known to mediate many aspects of plant adaptation to abiotic stresses [76]. ABA was shown to induce the expression of *CDSP34*, a potato gene highly similar to fibrillin and Arabidopsis *PGL35* [57]. The authors further showed that transcripts from the tomato *CDSP34* ortholog also accumulated after either ABA or dehydration treatments in wild-type plants and in *flacca* tomato mutants (impaired in ABA biosynthesis). In contrast, accumulation of the CDSP34 protein did not occur in the *flacca* background unless exogenous ABA was supplied, indicating that an ABA-dependant post-transcriptional mechanism controls the expression of *CDSP34*. In a recent report, Yang et al. [63] demonstrated accumulation of PGL35 after exogenous application of ABA. Moreover, using yeast two-hybrid and pull-down assays, a direct interaction between the transit peptide of PGL35 and ABI1 (a key player of ABA signaling) was shown, further implicating ABA in the post-transcriptional regulation of PGL expression [63]. *VTE1* transcripts also accumulated after 3 h treatment with 10 μM ABA (3.3-fold, Genevestigator, [37]). These observations suggest that ABA induces the accumulation of tocopherols both by up-regulating their synthesis and by increasing the volume of plastoglobules, their storage compartment.

The first, rate-limiting step of ABA biosynthesis is the cleavage of 9-*cis*-epoxycarotenoids (NCE) [77]. Interestingly, a NCE dioxygenase isoform (CCD4, At4g19170) was identified in the plastoglobule proteome [32, 36]. To address the possible involvement of CCD4 in ABA biosynthesis, future work will address whether the protein possesses a NCE cleavage activity. Arabidopsis *ccd4* null mutants may also prove useful tools to answer this question. Plastoglobules serve as storage compartments for quinones and tocopherols in chloroplasts. They might therefore represent "sensors" of the redox status of plastids. Taking this into consideration, the possible localization of (part of) ABA biosynthesis in plastoglobules may be rationalized. The allene oxide synthase, involved in the biosynthesis of jasmonic acid (JA) precursor OPDA, was identified in the plastoglobule proteome [32, 36]. As for ABA biosynthesis, plastoglobules in addition to containing lipid precursors may act as redox sensors for JA biosynthesis.

4
Common Features of Lipid Bodies in Plant, Fungal and Animal Cells

In plants, animals and microorganisms, lipid bodies have functions as diverse as energy storage, structural lipid storage, lipid transport and lipid metabolism (reviewed in [40]). However, common themes exist between lipid bodies which will be briefly discussed.

4.1
General Organization of Lipid Bodies

Lipid bodies consist of a hydrophobic core surrounded by a monolayer of amphiphatic lipids. Peripherally associated proteins are found in most types of lipid bodies and are more or less tightly bound to their surface. Such an architecture was proposed for chromoplast fibrils [23, 78], and chloroplast plastoglobules [29].

4.2
Association of Peripheral Proteins

Proteins associated with lipid bodies have diverse physicochemical properties and topologies, reflecting various modes of association with the lipidic structures. Hydrophobic sequences in several peripheral oil body proteins ensure their association with the lipids. For example, hydrophobic domains in perilipin were shown to be essential for targeting and anchoring the protein to lipid droplets in adipocytes [79]. In desiccation tolerant seeds, oil bodies are coated with oleosins. A central hydrophobic domain in the proteins, referred to as "proline knot motif", is essential for their association with oil bodies [80] (Fig. 2B). Interestingly, association of the hepatitis C virus core protein with cytosolic lipid droplets in mammalian cells requires a proline-containing domain similar to that of oleosins [81].

Although several lipid body proteins are characterized by hydrophobic domains, others lack large apolar regions. The sequence of adipocyte differentiation related protein (ADRP), for example, does not contain obvious lipid-binding motifs (hydrophobic domains or amphiphatic α-helices; [40]) and discontinuous stretches of the protein are necessary for targeting to lipid bodies [82]. PGL proteins also lack strongly hydrophobic domains and their association with plastoglobules may therefore rely on interactions with surface lipids [30]. Indeed, several PGLs were identified in 8 M urea chloroplast membrane extractions [83], consistent with peripheral association with plastoglobules.

Lipid body proteins with highly diverse properties have similar functions. They are generally thought to prevent coalescence of lipid bodies with neighboring lipophilic structures. Oil body coalescence was for instance observed in seeds from oleosin-deficient plants [84]. In addition, certain lipid body proteins including ADRP induce the formation and regulate the size of the lipid bodies [85, 86]. Several lines of evidence (see Sect. 2.2) suggest that PGLs may also be involved in regulating the morphology of plastid lipid bodies. However, future studies are needed to address the underlying molecular mechanisms.

4.3
Metabolic Functions

The identification of enzymes associated with plastoglobules indicates that chloroplast lipid bodies are metabolically active [32, 36]. Moreover, the observation that chloroplast, chromoplast and etioplast low-density fractions have different protein assortments [36] suggests that enzymatic functions of plastid lipid bodies are highly dynamic.

The analysis of the proteome from Chinese hamster ovary (CHO) K2 cell lipid droplets also revealed metabolic functions [87]. In addition to structural proteins (including ADRP), enzymes involved in the synthesis, storage, utilization, and degradation of cholesterol esters and TAGs were identified. Rab GTPases, as well as a GTPase activating protein (p50RhoGAP) were also detected, suggesting functions in membrane traffic and signaling. Enzymes in lipid metabolism were also shown to associate with yeast oil bodies [88, 89]. Interestingly, sterol-Δ^{24}-methyltransferase (Erg6), but not other enzymes from ergosterol biosynthesis was localized to lipid bodies, indicating shuttling of intermediates between the endoplasmic reticulum and the lipid bodies [40]. Tocopherol biosynthesis in plastids represents a similar situation, with VTE1 associated with plastoglobules and the other metabolic steps occurring at the envelope [90]. With the exception of a 11-β-hydroxysteroid dehydrogenase-like protein, a calcium binding protein (ATS1), as well as two proteins of unknown function, proteins identified in association with oleosomes were structural oleosins [91]. Analysis of the *Brassica napus* seed proteome identified in addition a putative short chain dehydrogenase/reductase and a protein similar to GDSL-motif lipase/hydrolases, suggesting functions in lipid metabolism [92]. In the studies by Jolivet et al. [91] and Katavic et al. [92], mature oilseeds were used for protein identification. Because the protein assortment of oleosomes may be dynamic, it would however be interesting to analyze oil body proteomes in germinating and maturing seeds.

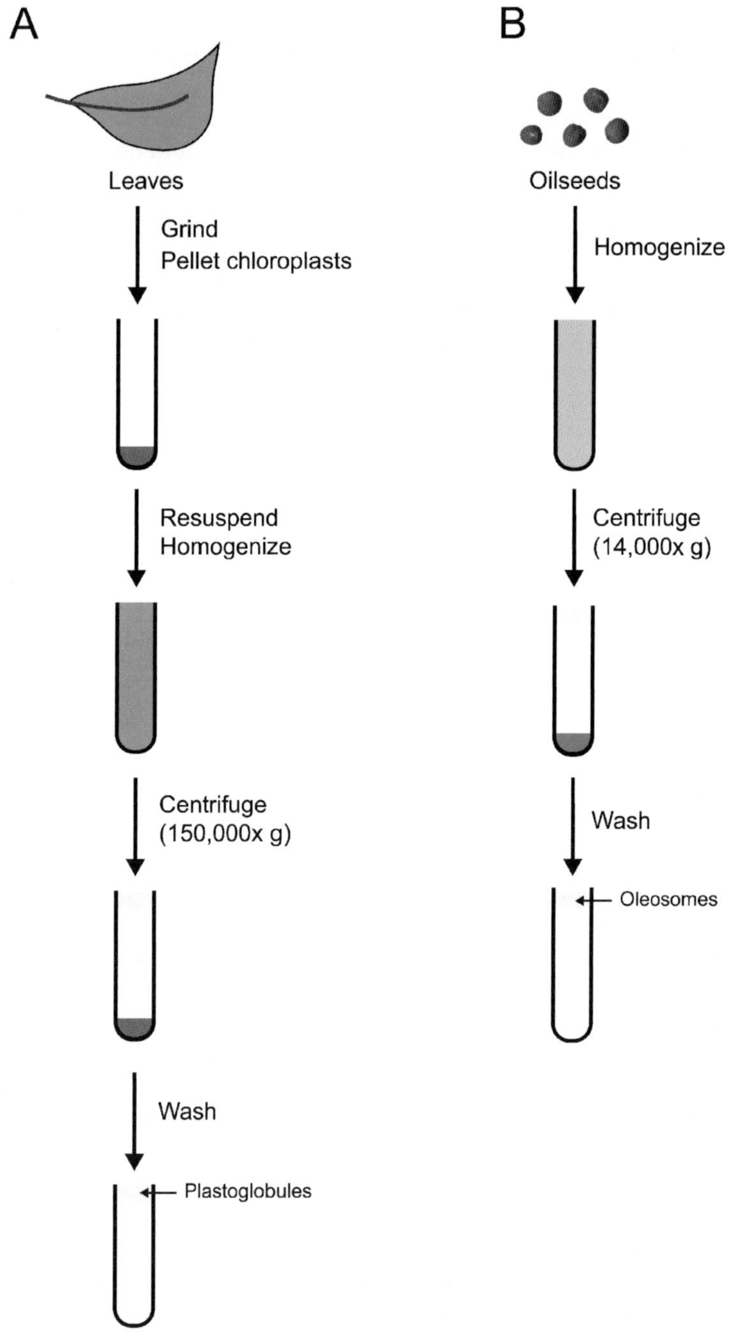

Fig. 3 Procedures for the purification of plant oil bodies. Purification schemes for chloroplast plastoglobules **A** adapted from [36] and oilseed oleosomes **B** as described in [95]. Both procedures rely on the low density of the lipid bodies

5 Potential of Plastoglobules for Bioengineering Applications

5.1 Purification of Plant Lipid Bodies

Purification procedures for plastoglobules [29, 32, 36, 93] and oleosomes [94, 95] rely on the low density of the lipid bodies (Fig. 3). In both systems, extracts containing plastoglobules or oleosomes are centrifuged. Membrane systems such as thylakoid or ER which contain high protein/lipid ratios sediment, while low-density lipid bodies accumulate on top of the supernatant and are recovered. While sucrose density gradients have been used in several studies to isolate plastoglobules [29, 32, 38, 93], Ytterberg et al. [36] have shown that a simpler procedure, similar to that used by van Rooijen et al. [95] for oleosome extraction, was also adequate for plastoglobule purification.

5.2 Plant Lipid Bodies as Purification Matrices for Recombinant Proteins

To date, proteins used in medicine as diagnostic reagents, drugs or vaccines are mostly produced by microbial or animal cell fermentation. These manufacturing systems allow highly controlled procedures but have disadvantages in term of cost and scalability. Pathogen contamination of animal cell cultures also represent an important safety issue. Plants stand as alternative systems for the production of recombinant proteins at lower costs ("molecular farming") [96, 97]. They allow large-scale production with accurate folding and assembly of protein complexes [98–101].

Important issues for industrial production of plant-derived recombinant proteins are extraction and purification [102]. Standard protocols include homogenization of plant biomass followed by chromatographic methods. However, developing cost-effective preliminary (or alternative) purification steps is of great interest since the high abundance of secondary compounds, especially in tobacco, is problematic for chromatographic procedures [97]. Plant cells contain roughly 30 000 different proteins whereas only nine different proteins were identified in oleosomes [91] and the proteome of plastoglobules consists solely of about 20 core components [32, 36]. Targeting recombinant proteins to lipid bodies and subsequent isolation of lipid bodies therefore represents an effective purification step.

The use of oleosomes as carriers for recombinant proteins was proposed by van Rooijen et al. [103] and the system is now being developed by SemBioSys (http://www.sembiosys.com/Index.aspx; [104]). The oil body-oleosin (Stratosome) system has for instance been applied to the production of biologically active human insulin [105]. A similar approach may be followed using plastoglobules derived from leaf crops such as tobacco. It would combine the high potential of plastids for recombinant protein accumulation [106, 107] with simple purification procedures. Moreover, sequestration of foreign proteins in lipid bodies may limit deleterious effects on chloroplast metabolism.

5.3 Plastoglobules as Sources of Tocopherol

Vitamin E is widely used in the industry, notably as a dietary supplement for animal nutrition as well as antioxidant in cosmetic and food preparations: about 40 000 tons of tocopherols were produced in 2002 [108]. Most tocopherol used in industrial applications is chemically produced. However, plants are interesting sources of tocopherol since (i) natural α-tocopherol has 1.5 more vitamin E activity than its synthetic racemic counterpart [66]. (ii) Plants represent renewable carbon sources, in contrast to fossil oils used for the chemical synthesis of tocopherol acetate. (iii) Additives from natural sources are becoming popular for consumers. To date, natural vitamin E is derived from soybean oil [108]. Considering the high tocopherol content of plastoglobules [32], it will be interesting to address the potential of plastoglobules as sources of tocopherols. Vitamin E could for example represent a valuable by-product in molecular farming applications.

6 Outlook

Plastoglobules have been observed in various plastid types for more than 40 years but have attracted little attention until recently. The conception of plastoglobules being passive storage compartments is now shifting toward a much more complex and dynamic view. The emerging picture implies plastoglobules in biosynthetic processes and in the ABA and JA signaling networks. The recent data on plastoglobules need to be connected with studies on other lipid bodies in yeast and animal systems where the same paradigm shift toward dynamic and metabolically active structures occurs. The publications reviewed in this work lay down a molecular basis for plastoglobule function, notably in stress protection but also raise many new questions: How are biosynthetic intermediates trafficked between plastoglobules and the other chloroplast membrane systems? How do plastoglobules integrate in ABA (and JA) regulatory networks? What is the function of the "unknown" plastoglobule proteins?

Evidence in the literature indicates a dynamic lipid composition of plastoglobules. The data are, however, scarce and better characterization of plastoglobule lipid composition in various developmental stages or environmental conditions will help to understand their physiological functions and may identify interesting compounds.

In the absence of structural information on PGLs, hypotheses were suggested regarding their mode of association with plastoglobules. Getting a better understanding of the mechanisms underlying protein targeting to plastoglobules may prompt new applications. We have drawn here the parallel between plastoglobules-PGLs and seed oil body-oleosin systems. Future work will address the potential of plastoglobules as targeting destinations for recombinant proteins. Also, simplification of the plastoglobule purification scheme should be considered, in order notably to reduce centrifugation steps. The identification of optimal conditions where plastoglobules strongly accumulate in plastids (e.g. upon stress treatment or during the ageing process) should also be part of future investigations.

References

1. Lopez-Juez E, Pyke KA (2005) Int J Dev Biol 49:557
2. Thomson WW, M. WJ (1980) Ann Rev Plant Physiol 31:375
3. Whatley JM (1978) New Phytol 80:489
4. McFadden GI (1999) Curr Opin Plant Biol 2:513
5. Bedard J, Jarvis P (2005) J Exp Bot 56:2287
6. Kessler F, Schnell DJ (2006) Traffic 7:248
7. Soll J, Schleiff E (2004) Nat Rev Mol Cell Biol 5:198
8. Hiltbrunner A, Bauer J, Alvarez-Huerta M, Kessler F (2001) Biochem Cell Biol 79:629
9. Greenwood AD, Leech RM, Williams JP (1963) Biochim Biophys Acta 78:148
10. Lichtenthaler HK, Sprey B (1966) Z Naturforsch 21:690
11. Sprey B, Lichtenthaler H (1966) Z Naturforsch 21b:697
12. Thomson W, Platt K (1973) New Phytol 72:791
13. Lichtenthaler HK (1968) Endeavor 27:144
14. Rey P, Gillet B, Romer S, Eymery F, Massimino J, Peltier G, Kuntz M (2000) Plant J 21:483
15. Austin JR 2nd, Frost E, Vidi PA, Kessler F, Staehelin LA (2006) Plant Cell 18:1693
16. Hoober JK, Maloney MA, Asbury LR, Marks DB (1990) Plant Physiol 92:419
17. van de Meene AM, Hohmann-Marriott MF, Vermaas WF, Roberson RW (2006) Arch Microbiol 184:259
18. Lichtenthaler HK, Tevini M (1970) Z Pflanzenphysiol 62:33
19. Legget Bailey J, Whyborn AG (1963) Biochim Biophys Acta 78:163
20. Lohmann A, Schottler MA, Brehelin C, Kessler F, Bock R, Cahoon EB, Dormann P (2006) J Biol Chem 281:40461
21. Steinmüller D, Tevini M (1985) Planta 163:201
22. Tevini M, Steinmüller D (1985) Planta 163:91
23. Deruere J, Romer S, d'Harlingue A, Backhaus RA, Kuntz M, Camara B (1994) Plant Cell 6:119
24. Newman LA, Hadjeb N, Price CA (1989) Plant Physiol 91:455
25. Pozueta-Romero J, Rafia F, Houlne G, Cheniclet C, Carde JP, Schantz ML, Schantz R (1997) Plant Physiol 115:1185
26. Vishnevetsky M, Ovadis M, Itzhaki H, Vainstein A (1997) J Biol Chem 272:24747

27. Vishnevetsky M, Ovadis M, Itzhaki H, Levy M, Libal-Weksler Y, Adam Z, Vainstein A (1996) Plant J 10:1111
28. Moriguchi T, Kita M, Endo-Inagaki T, Ikoma Y, Omura M (1998) Biochim Biophys Acta 1442:334
29. Kessler F, Schnell D, Blobel G (1999) Planta 208:107
30. Kim HU, Wu SS, Ratnayake C, Huang AH (2001) Plant Physiol 126:330
31. Ting JT, Wu SS, Ratnayake C, Huang AH (1998) Plant J 16:541
32. Vidi PA, Kanwischer M, Baginsky S, Austin JR, Csucs G, Dormann P, Kessler F, Brehelin C (2006) J Biol Chem 281:11225
33. Eymery F, Rey P (1999) Plant Physiol Biochem 37:305
34. Monte E, Ludevid D, Prat S (1999) Plant J 19:399
35. Laizet Y, Pontier D, Mache R, Kuntz M (2004) J Genome Sci Technol 3:19
36. Ytterberg AJ, Peltier JB, van Wijk KJ (2006) Plant Physiol 140:984
37. Zimmermann P, Hirsch-Hoffmann M, Hennig L, Gruissem W (2004) Plant Physiol 136:2621
38. Wu SS, Platt KA, Ratnayake C, Wang TW, Ting JT, Huang AH (1997) Proc Natl Acad Sci USA 94:12711
39. Hirschberg J (2001) Curr Opin Plant Biol 4:210
40. Murphy DJ (2001) Prog Lipid Res 40:325
41. Ghosh S, Mahoney SR, Penterman JN, Peirson D, Dumbroff EB (2001) Plant Physiol Biochem 39:777
42. Keskitalo J, Bergquist G, Gardestrom P, Jansson S (2005) Plant Physiol 139:1635
43. Tuquet C, Newman DW (1980) Cytobios 29:43
44. Kaup MT, Froese CD, Thompson JE (2002) Plant Physiol 129:1616
45. Guiamet JJ, Pichersky E, Nooden LD (1999) Plant Cell Physiol 40:986
46. Vishnevetsky M, Ovadis M, Zuker A, Vainstein A (1999) Plant J 20:423
47. Locy RD, Chang CC, Nielsen BL, Singh NK (1996) Plant Physiol 110:321
48. Bondada BR, Syvertsen JP (2003) Tree Physiol 23:553
49. Duret S, Bonaly J, Bariaud A, Vannereau A, Mestre JC (1986) Environ Res 39:96
50. Panou-Filotheou H, Bosabalidis AM, Karataglis S (2001) Ann Bot (Lond) 88:207
51. Merzlyak M, Solovchenko A, Pogosyan S (2005) Photochem Photobiol Sci 4:333
52. Ebel B, Rosenkranz J, Schiffgens A, Lutz C (1990) Environ Pollut 64:323
53. Oksanen E, Sober J, Karnosky DF (2001) Environ Pollut 115:437
54. Nilsen ET, Stetler DA, Gassman CA (1988) Am J Bot 75:1526
55. Munne-Bosch S, Alegre L (2002) Planta 214:608
56. Chen H-C, Klein A, Xiang M, Backhaus RA, Kuntz M (1998) Plant J 14:317
57. Gillet B, Beyly A, Peltier G, Rey P (1998) Plant J 16:257
58. Kuntz M, Chen HC, Simkin AJ, Romer S, Shipton CA, Drake R, Schuch W, Bramley PM (1998) Plant J 13:351
59. Langenkamper G, Manac'h N, Broin M, Cuine S, Becuwe N, Kuntz M, Rey P (2001) J Exp Bot 52:1545
60. Manac'h N, Kuntz M (1999) Plant Physiol Biochem 37:859
61. Pruvot G, Cuine S, Peltier G, Rey P (1996) Planta 198:471
62. Pruvot G, Massimino J, Peltier G, Rey P (1996) Physiol Plant 97:123
63. Yang Y, Sulpice R, Himmelbach A, Meinhard M, Christmann A, Grill E (2006) Proc Natl Acad Sci USA 103:6061
64. Bailey S, Thompson E, Nixon PJ, Horton P, Mullineaux CW, Robinson C, Mann NH (2002) J Biol Chem 277:2006
65. Munne-Bosch S (2005) J Plant Physiol 162:743
66. Schneider C (2005) Mol Nutr Food Res 49:7

67. Havaux M, Eymery F, Porfirova S, Rey P, Dormann P (2005) Plant Cell 17:3451
68. Sattler SE, Gilliland LU, Magallanes-Lundback M, Pollard M, DellaPenna D (2004) Plant Cell 16:1419
69. Collakova E, DellaPenna D (2003) Plant Physiol 133:930
70. Kanwischer M, Porfirova S, Bergmuller E, Dormann P (2005) Plant Physiol 137:713
71. Maeda H, Sakuragi Y, Bryant DA, Dellapenna D (2005) Plant Physiol 138:1422
72. Porfirova S, Bergmuller E, Tropf S, Lemke R, Dormann P (2002) Proc Natl Acad Sci USA 99:12495
73. Lichtenthaler HK, Peveling E (1967) Planta (Berl) 72:1
74. Thomas DJ, Thomas J, Youderian PA, Herbert SK (2001) Plant Cell Physiol 42:803
75. Lichtenthaler HK (1969) Ber Dtsch Bot Ges 82:483
76. Shinozaki K, Yamaguchi-Shinozaki K (2000) Curr Opin Plant Biol 3:217
77. Schwartz SH, Qin X, Zeevaart JA (2003) Plant Physiol 131:1591
78. Knoth R, Hansmann P, Sitte P (1986) Planta 168:167
79. Subramanian V, Garcia A, Sekowski A, Brasaemle DL (2004) J Lipid Res 45:1983
80. Hsieh K, Huang AH (2004) Plant Physiol 136:3427
81. Hope RG, McLauchlan J (2000) J Gen Virol 81:1913
82. Targett-Adams P, Chambers D, Gledhill S, Hope RG, Coy JF, Girod A, McLauchlan J (2003) J Biol Chem 278:15998
83. Kleffmann T, Russenberger D, von Zychlinski A, Christopher W, Sjolander K, Gruissem W, Baginsky S (2004) Curr Biol 14:354
84. Siloto RM, Findlay K, Lopez-Villalobos A, Yeung EC, Nykiforuk CL, Moloney MM (2006) Plant Cell
85. Fukushima M, Enjoji M, Kohjima M, Sugimoto R, Ohta S, Kotoh K, Kuniyoshi M, Kobayashi K, Imamura M, Inoguchi T, Nakamuta M, Nawata H (2005) In Vitro Cell Dev Biol Anim 41:321
86. Imamura M, Inoguchi T, Ikuyama S, Taniguchi S, Kobayashi K, Nakashima N, Nawata H (2002) Am J Physiol Endocrinol Metab 283:E775
87. Liu P, Ying Y, Zhao Y, Mundy DI, Zhu M, Anderson RG (2004) J Biol Chem 279:3787
88. Athenstaedt K, Zweytick D, Jandrositz A, Kohlwein SD, Daum G (1999) J Bacteriol 181:6441
89. Leber R, Zinser E, Zellnig G, Paltauf F, Daum G (1994) Yeast 10:1421
90. Soll J, Douce R, Schultz G (1980) FEBS Lett 112:243
91. Jolivet P, Roux E, D'Andrea S, Davanture M, Negroni L, Zivy M, Chardot T (2004) Plant Physiol Biochem 42:501
92. Katavic V, Agrawal GK, Hajduch M, Harris SL, Thelen JJ (2006) Proteomics 6:4586
93. Picher M, Grenier G, Purcell M, proteau L, Beaumont G (1993) New Phytol 123:657
94. Tzen JT, Peng CC, Cheng DJ, Chen EC, Chiu JM (1997) J Biochem (Tokyo) 121:762
95. van Rooijen GJ, Moloney MM (1995) Plant Physiol 109:1353
96. Joshi L, Lopez LC (2005) Curr Opin Plant Biol 8:223
97. Menkhaus TJ, Bai Y, Zhang C, Nikolov ZL, Glatz CE (2004) Biotechnol Prog 20:1001
98. Fischer R, Stoger E, Schillberg S, Christou P, Twyman RM (2004) Curr Opin Plant Biol 7:152
99. Ma JK, Barros E, Bock R, Christou P, Dale PJ, Dix PJ, Fischer R, Irwin J, Mahoney R, Pezzotti M, Schillberg S, Sparrow P, Stoger E, Twyman RM (2005) EMBO Rep 6:593
100. Ma JK, Chikwamba R, Sparrow P, Fischer R, Mahoney R, Twyman RM (2005) Trends Plant Sci 10:580
101. Ma JK, Drake PM, Christou P (2003) Nat Rev Genet 4:794
102. Giddings G, Allison G, Brooks D, Carter A (2000) Nat Biotech 18:1151
103. van Rooijen GJ, Moloney MM (1995) Biotechnology (NY) 13:72

104. Moloney M, Boothe J, Van Rooijen G (2003) US Patent 6509453
105. Nykiforuk CL, Boothe JG, Murray EW, Keon RG, Goren HJ, Markley NA, Moloney MM (2006) Plant Biotechnol J 4:77
106. Bock R (2001) J Mol Biol 312:425
107. Daniell H, Chebolu S, Kumar S, Singleton M, Falconer R (2005) Vaccine 23:1779
108. Valentin HE, Qi Q (2005) Appl Microbiol Biotechnol 68:436

Invited by: Professor Sautter

Adv Biochem Engin/Biotechnol (2007) 107: 173–205
DOI 10.1007/10_2007_050

Published online: 15 March 2007

Genetic and Ecological Consequences of Transgene Flow to the Wild Flora

François Felber[1] (✉) · Gregor Kozlowski[2] · Nils Arrigo[1] · Roberto Guadagnuolo[1]

[1]Laboratoire de Botanique évolutive, Institut de Biologie, Université de Neuchâtel, rue Emile-Argand 11, 2009 Neuchâtel, Switzerland
Francois.Felber@unine.ch

[2]Department of Biology and Botanical Garden, University of Fribourg, ch. du Musée 10, 1700 Fribourg, Switzerland

Abstract Gene flow from crops to wild relatives by sexual reproduction is one of the major issues in risk assessment for the cultivation of genetically engineered (GE) plants. The main factors which influence hybridization and introgression, the two processes of gene flow, as well as the accompanying containment measures of the transgene, are reviewed.

The comparison of risks between Switzerland and Europe highlights the importance of regional studies. Differences were assessed for barley, beet and wheat. Moreover, transgene flow through several wild species acting as bridge (bridge species) has been up to now poorly investigated. Indeed, transgene flow may go beyond the closest wild relative, as in nature several wild species complexes hybridize. Its importance is assessed by several examples in Poaceae. Finally, the transgene itself has genetic and ecological consequences that are reviewed. Transgenic hybrids between crops and wild relatives may have lower fitness than the wild relatives, but in several cases, no cost was detected. On the other hand, the transgene provides advantages to the hybrids, in the case of selective value as a Bt transgene in the presence of herbivores. Genetic and ecological consequences of a transgene in a wild species are complex and depend on the type of transgene, its insertion site, the density of plants and ecological factors. More studies are needed for understanding the short and long term consequences of escape of a transgene in the wild.

Keywords Risk assessment · Transgene · Genetically engineered plants · Bridge species · Switzerland

Abbreviations

GE genetically engineered
FOEN Swiss Federal Office for the Environment

1 Introduction

The gene transfer from crops into populations of wild relatives has become an important scientific and public issue since the development and cultivation of genetically engineered (GE) plants in the late 1980s [1]. The concerns related to the cultivation of GE crops, in particular those dealing with the possibility of transgene escape into the wild flora, have generated a multitude of studies on crop-to-wild gene flow [1–9]. While these studies have shown that such gene flow exists for almost all of the most important crops cultivated worldwide, only recently have new studies focused and are focusing on its ecological and genetic consequences [9–11]. Yet, in order to better assess the ecological and agronomic risks associated with the transgene flow to the wild flora, it is fundamental to understand the mechanisms and the consequences of such gene flow [12].

Studies on the existence of crop-to-wild gene flow have already been reviewed several times in the context of the cultivation of GE crops (e.g. [2, 5, 7]). A general overview on the factors influencing gene flow, and containment measures is presented here. Risk of gene flow has a geographical component and we focus in Sect. 2 on the particular case of Switzerland. While gene flow has been mostly investigated from crops to their closest wild relatives, further introgression may occur between wild species. The importance of such "bridge species" is explained in Sect. 3. Finally, the genetic and ecological consequence of transgene flow is evaluated in Sect. 4.

1.1
Factors Influencing Gene Flow

It is widely accepted that hybridization between two taxa depends on several key factors, such as their sympatry, the synchrony of their flowering periods, the existence of a common vector for the gametes, as well as their reproductive compatibility and the viability and fertility of the hybrids [12]. Generally speaking, gene flow between two taxa can thus be viewed as a two step process: (i) a first hybridization event, which leads to the production of first generation hybrids, followed by (ii) the introgression of part of the genome of one species into the other by successive backcrosses.

Hybridization depends mainly on straightforward conditions, such as the need for the plants to grow close to each other and the potential for the exchange of pollen. While the most obvious situation where both conditions are met is represented by cultivated fields where the wild relatives grow in close proximity, it is worth noting that crop plants growing as volunteers within fields of other crops or in other habitats represent additional contact zones between crops and wild relatives. In the case of GE crop, such a situation may lead to transgene escape to the wild flora. A notable example of this latter situation is that of rapeseed, which is extremely common to see in any kind of disturbed habitats even relatively distant from cultivated areas [6].

More generally, the establishment of feral crop populations in the agroecosystems, as well as outside the cultivated areas depends mainly on the crop features, such as seed dispersal by wind, water or animals, absence of dormancy, ripening period, persistence of seeds in the soil [13, 14]. Agricultural practices (harvesting period, crop rotation, till vs. no-till) as well as post-harvesting procedures (transportation), can also greatly influence the emergence of volunteer plants.

Finally, an additional potential source of transgenes is represented by first and subsequent generations of hybrids between GE crops and wild relatives, which can act as "genetic bridges" between the parental species [5, 9].

1.2
Factors Influencing Hybridization

Hybridization is influenced quantitatively by numerous factors, some of them depending on the characteristics of the plants, while others are more related to the environment. Hybridization is frequent in perennial species and especially for outcrossing and clonal plants [2], as the produced hybrids can subsist clonally even in the case of reduced fertility.

Pollen vectors play a major role, at least on the distance at which hybridization can take place. For instance, maize pollen is known to be particularly heavy and intraspecific gene flow at distances greater than 50 meters is unlikely [15]. In contrast, other wind-pollinated species can show large distance pollen dispersal events. Watrud et al. [16] discovered intraspecific hybrids of *Agrostis stolonifera* 21 km from the pollen source.

While it is obvious that topography influences winds, a flat land favoring the pollen flow over long distances, it is worth mentioning that microtopography seems to have also an impact on the behavior of pollinator insects, by hiding or making more visible potential pollen sources and sinks. However, predictions on the pollen movements seem more complicated in the case of insect pollinated species. Different experimental and modeling studies on the distance at which rapeseed pollen could produce hybrids, generated indeed inconsistent results [17, 18], because these results depend indirectly on the factors influencing the activity of bees [19].

Repeated contacts with crop populations are known to accelerate the introgression process [20]. However, hybridization as well is positively correlated with the frequency and the extent of the contact zones between crops and their wild relatives. Indeed, feral populations or individual volunteer crop plants will not only increase the area of contact, but also increase the potential for the overlap of flowering periods. For instance, while in central Europe fields of rapeseed usually flower simultaneously in May, it is common to observe volunteers flowering from June till late October.

1.3 Factors Influencing Introgression

Most factors influence both hybridization and introgression. While successful introgression is achieved when genes from one taxon are fixed in another one, several hybrid generations and parental individuals can be involved in the process. All of these individuals and generations can coexist and exchange genes simultaneously for many years [5].

Fitness of hybrids is essential to successful introgression. Moreover, independent of the pollen vector, the intensity and symmetry of pollen flow will determine both the direction of hybridization and the speed of introgression in a sink population. Fixation of genes is known to occur more rapidly in small populations, which are also more prone to act as a pollen sink [21, 22].

Both hybridization and introgression are facilitated in genetically close species, such as crop and prickly lettuce (D'Andrea et al., unpublished) or crop and wild sunflower [23], rather than between more distantly related species like rapeseed and wild radish [24]. The actual introgression of crop genes into the genome will depend greatly on the existence of pre- and postzygotic barriers, which strongly depend on factors linked to the evolutionary divergence between the crop and its wild relative, the incompatibilities being generally higher between genetically distant taxa and lower between closely related taxa.

Genetic barriers acting against hybridization between species are considered by several authors as "semi-permeable" [25, 26]. Individual genes or specific genome regions may be transferred during introgression processes, rather than entire genomes [5]. Moreover, genes from one species may not be uniformly transmitted to another via introgressed generations, as selection does not act homogeneously within genomes.

A factor that influences specifically introgression, rather than hybridization, is the observation that beneficial or neutral traits will be preferentially introgressed, compared to detrimental genes. For example, silenced genes can be kept in recipient genomes, until they are eliminated by genetic drift [23]. Additionally, several linked genes may be transferred together, especially if such complexes carry positively selected genes.

The situation is more complex in polyploids where multiple copies of genes make genetic interactions even more complicated. Moreover, related polyploid species often share only part of the genome (e.g *Triticum aestivum* and *Aegilops cylindrica*, *Brassica napus* and *B. campestris*) and introgression from one species to the other is easier for genes located on the homologous chromosomes, than for genes located in the homeologous ones.

1.4 Containment of Transgenes

One clue which arises from the existing studies on crop-to-wild gene flow is that hybridization between most crops and their wild relatives cannot be avoided [4]. Therefore, if the goal is to impede the transfer of transgenes to the wild flora, gene flow has to be stopped at its source. For this purpose, several strategies, each possessing advantages and drawbacks, have been proposed (most are reviewed in [5]), which are mostly linked to the mechanisms and factors influencing introgression presented above.

Since physical barriers, such as isolation by distance or hedge rows bordering fields appeared rapidly to be inefficient, genetic barriers based on the breeding systems of the crops were investigated. One of the first ideas was to decrease or completely block gene flow via pollen, by favoring apomixis. However, many apomictic species preserve low to moderate sexual seed production, and moderate or high levels of pollen [14].

It was thus suggested to induce male-sterility in GE crops. This system was applied to commercialized *Brassica napus* varieties [27]. In this rapeseed variety, the transgenic construct is induced by a *tapetum*-specific promoter, and produces a cytotoxin (*barnase*). Only anthers express the lethal transgene, which leads to the destruction of the mother cells of pollen. However, male sterility does not prevent the formation of hybrids when wild relatives act as paternal parent, like in the case of bolting beets in south Europe [4]. Moreover, these two strategies can only prevent gene flow by pollen, while they have no effect on gene flow by seeds.

It was subsequently suggested to insert transgenes in genomic regions, which have no or reduced mobility. As mentioned previously, genomes are not uniformly transmitted, and some regions are more "mobile" than others [25, 28]. Targeting gene insertion in regions poorly transmitted should decrease the probability of gene escape. However, in order to be efficient, this strategy has to be developed on a case-by-case basis, and introgressive patterns on all possible wild relatives of each crop should be known. A similar idea was proposed for polyploid species, where genomes non-shared by wild relatives could be chosen as insertion sites of transgenes (e.g. [29, 30]). However, recombination events between non-homeologous genomes were observed in wild x crop hybrids involving *Brassica napus* [31], and *Triticum aestivum* [32, 33].

Another proposition was to insert transgenes in the DNA of mitochondria or chloroplast, as organellar DNA is usually maternally transmitted, and should not be carried by pollen grains in Angiosperms [34]. However, paternal inheritance of chloroplasts has already been observed (reviewed by [35]). For instance, transfer of genes from organelles to nucleus occurs at a low frequency in tobacco, as one pollen grain out of 16 000 carries cytoplasmic genome elements in its nucleus [36]. As for the strategies presented above, gene flow via seeds is not prevented.

Therefore, so-called "seed suicide" techniques were proposed (see [12] for a review). In these plants, the transgenic construct induces the production of lethal protein or blocks physiological functions during seed maturation, which makes it impossible for the seeds to germinate, but without disturbing albumen differentiation. However, producing non-germinating seeds would impede farmers from sowing part of their harvest, which is a highly controversial issue from an ethical point of view.

Another recent technique consists of the chemically induced removal of transgene from pollen cells during the gametogenesis. The transgene is flanked by specific sites (*lox*), which allows its removal by a site-specific recombinase (Cre). The recombinase is coded by the transgene and expressed after induction [37]. Recombinase-based techniques present currently two major drawbacks: the controlling system has to be activated by an external signal, that is the application of tetracycline, and basically every single cell involved in the sexual reproduction of the crop should be treated.

Finally, post-hybridization and fitness-based strategies were also suggested to avoid the spread of hybrid derivates in the environment. The idea is to lower the fitness of these plants by linking the transgene with traits which are neutral or beneficial in an agricultural context, but detrimental in the wild. The genes responsible for traits such as dwarfing, loss of dormancy or non-shattering of seeds were proposed as suitable loci to place transgenes [5]. However, there are at least two serious drawbacks in this strategy. First, the current technology does not allow placing of the transgenic construct in a precise location. Second and probably more important, most of these so-called deleterious traits are recessive loss-of-function alleles related to the domestication of crops [13]. These alleles would thus not be expressed in first generation hybrids with a wild plant, because of the presence of the dominant wild allele in their genome. In further generations, the deleterious allele would only be expressed in homozygous individuals, which would strongly reduce its capability to lower the fitness of these plants. Moreover, if the hybrids are fertile, this strategy would not prevent them acting as a genetic bridge and pollinating the wild parent [11].

Alternatively, this strategy could have a good efficacy when the transgene is coupled within the transgenic construct itself with one or two mutant genes conferring an ecological disadvantage (transgenetic mitigation, [38]), such as dwarfing, as demonstrated in tobacco introgressants [39].

2
Gene Flow between Cultivated Plants and Wild Relatives: the Case of Switzerland

Risks related to transgenic plants are often investigated on a worldwide scale and several reviews have focused on this topic. Nevertheless, a regional perspective is necessary because crops vary among countries, wild species have often a limited geographical range and floras composition changes geographically. Consequently, the distribution of crops and their ability to cross with their wild relatives vary regionally. Moreover, the genetic characteristics of a wild species, as for example its ploidy level, may vary according to their geographical range and can influence largely their ability to hybridize. This is illustrated for example by tetraploid alfalfa, *Medicago sativa* ($2n = 32$ chromosomes). In Switzerland, its wild relative, *Medicago falcata*, is tetraploid and has the same chromosome number ($2n = 32$) except in Unterengadin, where it is diploid ($2n = 16$). Hybrids between the two species, *M. x varia*, are found frequently where both species are tetraploid, but are, on the contrary, very rare in the range of the diploid *M. falcata* [40]. Risks of gene flow are consequently much lower in Unterengadin than in the other areas of Switzerland.

Consequently, the results of one country cannot be necessarily generalized to another country without further investigations. This is particularly true for Switzerland, where topography strongly influences the distribution of wild species and constrains agriculture. Its landscape typically illustrates that risks may vary from one area to the other.

Distribution of wild relatives may also vary in time. For example, global change, including both the global warming and the increase in disturbance as a consequence of human activity, has led to the northern expansion of several Mediterranean species. Similarly, change in agricultural practices may influence the contact zones between crops and their wild relatives, and consequently influence greatly the risks. Therefore, monitoring over a long term the wild flora and the agricultural areas is necessary in order to evaluate the risks on a regional perspective. Switzerland has voted on November 27, 2005 a moratorium of 5 years on the outdoor cultivation of GE organisms for commercial purposes. Probabilities of large-scale cultivation of transgenic plants are therefore low. Nevertheless, political changes may occur rapidly and therefore, assessment of potential risk of gene flow from crops to wild relative is necessary with a Swiss perspective.

The Swiss Federal Office for the Environment (FOEN) granted a study on risk assessment which focused on the main cultivated plants of Switzerland. For each of them, bibliographical data were collected on the crop and on most of the wild relatives (Table 1). From this, the risks of transfer of transgene to conventional varieties and to the wild or naturalized flora were evaluated (Table 2).

Risks of gene flow were never null between cultivars, as all crops reproduce sexually. Risks were evaluated as null for the wild flora when the crop produces no feral populations and no wild relative exists in Switzerland. It was low to medium in the case of autogamy, or of harvest before flowering (as for lettuce, out of the seed production areas). Risk was considered as high for all allogamous species, those forming spontaneous or subspontaneous populations, or possessing wild relatives that hybridize readily with the crop.

Risks may be examined in different perspectives. For a monitoring program, the priority is to examine commercialized transgenic crops. Prior to the authorization of outdoor cultivation, it is important to evaluate on one hand the risks of contamination of non-transgenic cultivation, and on the other hand those of gene flow to the wild flora.

Table 1 Characteristics collected for the crops and its wild relatives

Common for the crops and its wild relatives	Specific to the cultivated plant	Specific to the wild relatives
Latin name	Extent of cultivation	Ecology
Vernacular names	Feral populations	Hybridization with the crop
Chromosome number	Frequent transformations which have led to a request for a field trial	Hybridization with other wild relatives
Pollen dispersal	Recent transformation	Category of threat according to the Swiss Red List [104]
Breeding system	GE field cultivation	Stability of the distribution
Longevity	Commercialization	
Levels of vegetation		

Table 2 Summary of risks for the main crops cultivated in Switzerland

Name		Crop to crop gene flow*	Gene flow with spontaneous and naturalized flora*		Commer-cialization**	Main transformation
			Suisse	Europe		
Poaceae						
Agrostis stolonifera L.	Creeping bentgrass	++	++	++	pending	Herbicide tolerance, agronomical properties
Avena sativa L.	Oats	+	++	++	FT	Virus Resistance (virus BYDV)
Cynodon dactylon (L.) Persoon	Bermuda grass	++	+	+	FT	Agronomical properties, herbicide tolerance
Festuca arundinacea Schreber s.l., *F. pratensis Hudson s.l.*	Fescue	++	++	++	FT	Product quality, fungal resistance
Hordeum vulgare L.	Barley	+	0	+	FT	Product quality, fungal resistance
Lolium perenne Hudson s.l., *L. multiflorum* Lamarck	Ryegrass	++	++	++	FT	Product quality
Poa pratensis L.	Smooth meadow-grass	++	++	++	FT	Herbicide tolerance
Triticum aestivum L., *Triticum spelta* L.	Wheat	+	+	++	FT	Herbicide tolerance, fungal resistance
Zea mays L.	Maize	++	0	0	Com	Insect resistance, herbicide tolerance

Table 2 (continued)

Name		Crop to crop gene flow*	Gene flow with spontaneous and naturalized flora*		Commer-cialization**	Main transformation
			Suisse	Europe		
Rosaceae						
Fragaria x ananassa	Strawberry	+	+	+	FT	Fungal resistance, herbicide tolerance
Malus domestica Borkh.	Apple	++	++	++	FT	Product quality (fruit quality), insect and bacterial resistance
Prunus avium L.	Cherry	++	++	++	FT	Modification of metabolism
Prunus domestica L.	European plum	++	++	++	pending	Virus resistance
Pyrus communis L.	Pear	++	++	++	FT	Product quality (fruit maturation)
Rubus idaeus L.	Raspberry	++	++	++	FR	Virus resistance, product quality
Asteraceae						
Cichorium intybus L.	Chicory	++	++	++	Com	Agronomical properties (male sterility), herbicide tolerance
Helianthus annuus L.	Sunflower	++	+	+	FT	Fungal resistance, insect resistance
Lactuca sativa L.	Lettuce	+	+	+	FT	Herbicide tolerance, product quality

Table 2 (continued)

Name		Crop to crop gene flow*	Gene flow with spontaneous and naturalized flora*		Commer-cialization**	Main transformation
			Suisse	Europe		
Fabaceae						
Glycine max L.	Soybean	+	0	0	Com	Herbicide tolerance, product quality
Medicago sativa L.	Alfalfa	++	++	++	Com	Herbicide tolerance, product quality
Pisum sativum L.	Pea	+	+	+	FT	Herbicide tolerance, virus resistance
Solanaceae						
Lycopersicon esculentum Miller	Tomato	+	0	0	Com	Product quality, insect resistance
Nicotiana tabacum L.	Tobacco	++	0	0	Com	Product quality, virus resistance
Solanum tuberosum L.	Potato	+	0	0	Com	Insect resistance (doryphore), product quality
Other families						
Beta vulgaris L.	Beet	+	0	++	Com	Herbicide tolerance, virus resistance
Brassica napus L.	Rapeseed	++	++	++	Com	Product quality (oil quality), herbicide tolerance
Brassica rapa L.	Rape	++	++	++	FT	Insect resistance (lepidopters), herbicide tolerance
Cucumis melo L.	Melon	++	0	0	FT	Virus resistance, product quality (fruit ripening)

Table 2 (continued)

Name		Crop to crop gene flow*	Gene flow with spontaneous and naturalized flora* Suisse	Europe	Commer-cialization**	Main transformation
Cucumis sativus L.	Cucumber	++	0	0	FT	Virus resistance, agronomic properties (salt tolerance)
Cucurbita pepo L.	Squash	++	0	0	Com	Virus resistance
Daucus carota L.	Carrot	++	++	++	FT	Fungal resistance (*Alternaria* tolerance), product quality
Dianthus caryophyllus L.	Carnation	++	0	0	Com	Product quality (colors modification)
Osteospermum ecklonis (DC) Norl	Cape Daisy	++	0	0	FT	Metabolism modification
Picea abies (L.) Karsten	Norway Spruce	++	++	++	FT	Gene marker
Pinus sylvestris L.	Scots pine	++	++	++	FT	Gene marker, forestry performance
Populus alba x tremula, *Populus* sp.	Poplar	++	+	+	FT	Herbicide tolerance, forestry performance
Vitis vinifera L., *Vitis labrusca* L.	Grape	++	+	+	FT	Fungal resistance, virus resistance

* Evaluation of risks: 0 = no risk, + = low or medium risk, ++ = high risk
** Commercialization: FT = field tests have been carried out; pending = commercialization of transgenic varieties is pending; Com = transgenic varieties are commercialized

2.1 Priority Species for a Monitoring Program

Eleven cultivated species in Switzerland possess commercialized transgenic varieties elsewhere in the world. Six of them present a high risk of gene flow with other cultivars (alfalfa, carnation, chicory, maize, rapeseed and squash). Others represent a lower risk as they are harvested before flowering, such as beet, or because seed do not mature in the regions, such as potato for example. Moreover, crops with an autogamous breeding system such as soybean, tobacco or tomato also present a lower risk of gene flow.

2.2 Potential Risks for the Contamination of Non-Transgenic Crops

Only crops which have commercialized GE varieties are mentioned below. The higher risks originate from the six allogamous species mentioned above. Medium risks are characteristics from either autogamous species with partial allogamy, or those which are not producing fruits in traditional practices (beet, potato). Low risks exist for plants that do not flower, when vegetative parts are collected. Such cultivation necessitates a good management and strict control, in order to avoid any loss of seeds or unintended flowering. Some of the species mentioned above belong to that category, depending on their use.

2.3 Potential Risks for Gene Flow to the Wild or Naturalized Flora

High risk characterizes crops and wild relatives with no or low reproductive barriers, as for oilseed rape and creeping bentgrass. For example, escape of transgenic creeping bentgrass (*Agrostis stolonifera* L.) in non-agronomic areas was observed in the USA [41]. Medium risk occurs if the hybrid is partially fertile and introgression is possible. No commercialized transgenic crops belong to that category: pea, poplar, strawberry, sunflower and wheat. Some cultivated species have no wild relative in Switzerland; this is for example the case for beet, carnation, maize, melon, potato, soybean, tobacco, tomato and squashes. Consequently, they do not represent a genetic threat for the natural flora, even if containment measures are needed to avoid crop to crop gene flow.

2.4 Particularity of the Swiss Flora

Table 2 reveals that, for some species, different risks were assessed between Switzerland and Europe. Barley present no risk for Switzerland, as no ancestor grows in this country [42], while in the eastern Mediterranean to Iran and West Central Asia, hybridization occurs readily with *Hordeum spontaneum* [4]. Wheat presents also a lower risk in Switzerland, where only *Ae. cylindrica* forms durable populations, contrasting with the Mediterranean area where several wild relatives of *Aegilops* are frequent. Finally, beet presents no risk of outcrossing with the wild flora because its wild relative *Beta vulgaris* subsp. *maritima* is absent in Switzerland, while it is present close to the Atlantic coast and along the Mediterranean boarder.

3 The Importance of Bridge Species

Historically, the term "bridge species" has been used to designate wild plant species which could act, through artificial or natural hybridization, as a genetic bridge between wild relatives and closely related cultivated plants. Figure 1 shows that there are potentially three possible directions of the gene flow through bridge species: (1) wild-to-crop bridges, (2) crop-to-wild bridges and (3) wild-to-wild bridges.

The wild-to-crop bridges have been used by humans since millennia and are still used by breeders for the introduction of desirable traits from wild relatives into crops [43]. As discussed above, the development of GE crops has brought much more attention to the gene flow the other way around, that is between cultivars/crops and their wild relatives [2, 5, 7]. Surprisingly, the potential further spread of transgenes to other wild relatives via wild-to-wild

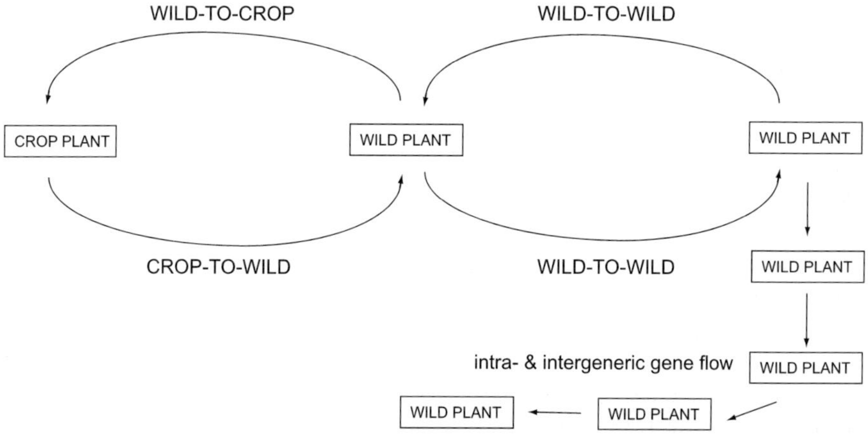

Fig. 1 Bridge species and directions of gene flow in crop-wild hybrid complexes. The potential spread of transgenes into wild populations via wild-to-wild bridges and further introgression has been poorly investigated

bridges and so-called stepping-stones introgression has been rarely if ever studied in details in the GE plants context. Thus, after a short description of the first well studied and described gene flow direction, more attention will be devoted to the still weakly explored subject of wild-to-wild bridges.

3.1 Wild-to-Crop Bridges

The term "bridge species" designates here wild relatives of cultivated plants which are used during artificial and/or natural hybridization procedures for crop improvement to circumvent some experimental or environmental constrains (Fig. 1). The ability to transfer genes between related plant species has been a great benefit in the improvement of cultivars for disease resistance, insect resistance, and/or end-use quality. This has been especially true in allopolyploid crops where there are multiple species that can act as donors. The best documented examples come from studies of gene transfer from wild species to wheat (*Triticum aestivum*). Romero et al. [44] obtained for example the transfer of a cereal cyst nematode resistance gene from *Aegilops triuncalis* (donor) to hexaploid wheat using bridge species *T. turgidum*. Fernandes et al. [43] transferred to wheat stem and leaf rust as well as powdery mildew resistance from *Ae. squarrosa* (donor) through hybridization with *T. durum* (bridge species). Such methods imitate in fact the ancient hybridization events, which happened during evolution and domestication of some crop plants, e.g. the hexaploid wheat. This bridge species method with development of intermediate natural or artificially synthesized amphiploid hybrids is one of the available procedures to facilitate gene flow between wild relatives and crop. It has been used for many decades not only for wheat cultivars [45] but also for many other crops (e.g. *Brassicaceae* [46], *Gossypium* sp. [47], *Cucumis* sp. [48]). However, for numerous plant groups such approaches are very laborious and/or have low or no success (e.g. for some *Solanum* sp. [49]).

3.2 "New Old Issue": Wild-to-Wild Bridges and Stepping-Stones Introgression

The hybridization and introgression between wild plants is a very well known phenomenon. Ellstrand et al. [2] estimated that there are more than 1000 well studied and published examples of spontaneous plant hybridization. Although at generally low frequencies and over long periods of time, genes (and thus also transgenes) can be spontaneously introgressed between different wild species [5]. It is therefore surprising that there are practically no detailed studies and exhaustive reviews on the importance of wild-to-wild hybridizations and wild-to-wild bridges in the context of the transgene flow and GE crops (Fig. 1).

How common are the natural hybridization processes between wild plant taxa? To answer this question we have to remember that there are two possible outcomes of hybridization [1]. The first outcome is the present ongoing introgression. Mallet [50] based on fundamental work on hybrid flora of the British Isles by Stace [51] estimated that at least 25% of all wild plant species are able to hybridize spontaneously and/or are involved in ongoing introgression processes with other wild species. Ellstrand et al. [52] using the same data set concluded that up to 34% of families and 16% of all genera in Great Britain have at least one reported hybrid. Additionally, there are many very well-studied genera with numerous closely related species producing hybrid swarms, such as *Salix* [53], *Quercus* [54] or *Eucalyptus* [55]. Rieseberg [56], based on the calculations of Ellstrand et al. [52], concluded that we could expect a worldwide total of 27 500 hybrid combinations among all Angiosperms. He added however, that it could be strongly underestimated since many regions, especially the tropics, are weakly explored and documented as far as their hybrid flora is concerned.

The second result of hybridization is the ancient and present speciation [56–58]. Indeed, in many families and genera, polyploidization and hybridization were the main mode of speciation and diversification. Ellstrand et al. [2] based on the summarizing works of Grant [59] and Arnold [60] concluded that more than 70% of plant species originated from hybrids.

As a consequence of these two well-documented hybridization outcomes, it has been often stated that the natural interspecific and even intergeneric hybrid formation is ubiquitous and uniform among higher plants [61, 62] or even the rule rather than the exception [63]. However, Ellstrand et al. [52] demonstrated clearly that the spontaneous hybridization is non-randomly distributed among systematic plant groups. By analyzing five biosystematic floras from Europe, North America and the Hawaiian Islands they showed that certain phylogenetic groups are predisposed for hybridization. To the most important hybrid families in practically all analyzed regions belong such crop-plant families as Poaceae, Asteraceae, Rosaceae and Fabaceae. Ellstrand et al. [2] enumerated 13 of the most important food crops grown for human consumption. Among them seven belong to Poaceae (*Eleusine, Hordeum, Oryza, Saccharum, Sorghum, Triticum, Zea*), and three to Fabaceae (*Arachis, Glycine, Phaseolus*). Hybridization seems therefore to be concentrated in a relatively restricted fraction of families and/or genera. Moreover, many members of these highly hybridizing families have been genetically modified and mainly possess numerous wild relatives. Table 3 lists all major plant genera and families of European flora containing crop plants with reported genetic transformation and/or with GE species used for field trials. It shows how many wild relatives of transformed crops could be found in Europe and which of those taxa possess the highest ability for complex hybridization. Here again the family of Poaceae has the most important potential for wild-to-wild bridge formation. Numerous of its mem-

Table 3 Number of wild relatives in European flora of the most important crop plants used for genetic transformation (only genera with GE members commercialized or used for field trials are listed) and estimation of their potential for wild-to-wild bridge formation. Symbols: – hybridization not observed, + hybridization observed but rare, ++ hybridization frequent, +++ hybridization extremely frequent (based on [51, 71], for *Lolium*, *Festuca* and *Poa* see also [112–160])

Genus	Nb. of wild (naturalized) species in Europe*	Potential for wild-to-wild bridge formation in Europe		
		intrageneric	intergeneric	hybridizing with
***Poaceae*:**				
Agrostis	24 (1)	+++	++	*Polypogon, Calamagrostis*
Avena	10 (2)	++	–	–
Cynodon	1	–	–	–
Festuca	165	+++	++	*Lolium, Vulpia*
Hordeum	8 (1)	++	++	*Agropyron, Elymus*
Lolium	5	+++	++	*Festuca*
Poa	43 (1)	+++	–	–
Triticum	3	+++	+++	*Aegilops, Elymus, Secale*
Zea	0	–	–	–
***Rosaceae*:**				
Fragaria	4 (1)	+	–	–
Malus	6	++	++	*Pyrus, Sorbus*
Prunus	19 (2)	++	–	–
Pyrus	11	++	++	*Malus, Sorbus*
Rubus	c. 75 (c. 3)	+++	–	–
Asteraceae:				
Cichorium	3	–	–	–
Helianthus	3 (7)	+	–	–
Lactuca	15	++	–	–
***Fabaceae*:**				
Glycine	0	–	–	–
Medicago	35 (2)	+	–	–
Pisum	1	–	–	–
***Solanaceae*:**				
Lycopersicon	0	–	–	–
Nicotiana	3(4)	–	–	–
Solanum	3 (9)	+	–	–
Other families:				
Beta	5	+	–	–
Brassica	20	+++	+++	*Raphanus, Sinapis, Diplotaxis, Hirschfeldia, Eruca, Erucastrum*
Cucumis	(1)	–	–	–
Cucurbita	0	–	–	–

Table 3 (continued)

Genus	Nb. of wild (naturalized) species in Europe*	Potential for wild-to-wild bridge formation in Europe		
		intrageneric	intergeneric	hybridizing with
Daucus	10	+	–	–
Picea	2 (c. 8)	–	–	–
Pinus	13 (c. 13)	–	–	–
Populus	4 (c. 6)	+++	–	–
Vitis	1 (c. 9)	+	–	–

* wild: native species including cultivated species capable of formation of weedy subspontaneous populations; naturalized: non-native species naturalized in Europe.

bers form easily both intra- and intergeneric hybrids. The only groups which could be compared with Poaceae are some genera of the family Rosaceae (mainly fruit trees) and the very well-known *Brassica* coenospecies complex.

It is worth mentioning that detailed studies and surveys on natural hybridization in wild taxa are extremely difficult. Abbott [64] and Ellstrand et al. [52] pointed out that the main limitation is the scarcity of modern biosystematic floras containing complete ecological, evolutionary and genetic information needed for such surveys. Additionally, the documentation of hybridization and introgression faces several methodological and theoretical difficulties. There are many methods used in identifying hybrids ranging from relatively simple morphological measurements to complex molecular and phylogenetic analyses. However, the majority of them, if not all, suffer from the fact that there are multiple explanations for the morphological and/or molecular intermediacy of a given hybrid candidate taxon [57]. The morphological similarity for example could be simply a result of convergent evolution. Martinsen et al. [26] concluded that the hybrid detection based on morphological characteristics is additionally constrained by backcrosses, since it is known that a backcrossed hybrid often resembles the parental species. The development of molecular genetic markers has facilitated studies of hybridization and allowed one to detect even very low levels of introgression. Additionally, it is possible with the molecular markers to track both the nuclear and cytoplasmic gene flow. However, the presence in one individual of molecular markers from two different species could be explained not only through recent hybridization but also due to shared ancestral characters (symplesiomorphy [26, 57]). The differentiation between contemporary versus ancient introgression is difficult and has been studied in only a few taxa [26, 65].

3.3 *Poaceae*: Example of a Biologically Predisposed Family for Wild-to-Wild Bridge Formation

Table 3 and detailed comparative studies mentioned above [2, 51, 52] demonstrate clearly the enormous potential of the family *Poaceae* for the wild-wild hybridization processes. The importance of the *Poaceae* as an object of research reflects their ecological and biogeographical success as well as their enormous economic value [66]. They occupy almost every habitat around the world, often being the dominating organisms [67]. The *Poaceae* comprises about 10 000 species and between 600 and 900 genera [68, 69]. In addition to that, the family contains a very high percentage of species and cytotypes of polyploidy origin. More than 80% of grass species have undergone polyploidy which represents the highest percentage in Angiosperms. Such a high level can be explained by successive regressions and extensions of the ranges which would favour secondary contact zones between related taxa, their hybridization and their subsequent polyploidization [70].

According to Wipff [71] one of the most important grass groups being currently used in genetic transformations are the forage grasses as well as grass species used for turf and erosion control. Furthermore, Wipff gives four main reasons why this group is particularly at risk of spreading transgenes: (1) they have undergone relatively little domestication; (2) they have usually numerous wild relatives; (3) they grow often in sympatry with these; (4) they can grow as weeds outside cultivated areas or in other crop cultures. To this grass group belong such common and species-rich European and North American genera as *Lolium, Festuca, Poa* and *Agrostis*. In the United States not less than 187 field tests were carried out between 1993 and 2006 with transgenic *Agrostis stolonifera*, 36 with *Poa pratensis*, 26 with *Festuca arundinacea*, 17 with *Cynodon dactylon* and 6 with *Lolium perenne* [72]. All mentioned species possess numerous wild relatives in Europe (Table 3) and are capable of hybridizing easily with them (e.g. *Festuca* with ca. 165 species, numerous subspecies and swarms of hybrids in Europe).

Figure 2 gives an example of intrageneric wild-to-wild hybrid complexes in *Poa*. Genus *Poa* contains approximately 43 species in Europe (Table 3) and 300 species worldwide. Intergeneric hybridization is extremely common and results in serious classification difficulties [71, 73]. In *Poa pratensis*, which absorbed genomes from many different taxa, it is even impossible to trace its ancestors [71, 74]. It was shown additionally that F1 hybrids between different *Poa* species can be completely fertile [75]. Figure 2 shows that almost 1/3 of all European *Poa* species are able to hybridize. They have mainly sympatric distribution even at a local level and have similar phenology. It is additionally very probable that more detailed studies would reveal much higher levels of intrageneric hybridization between members of the genus *Poa*.

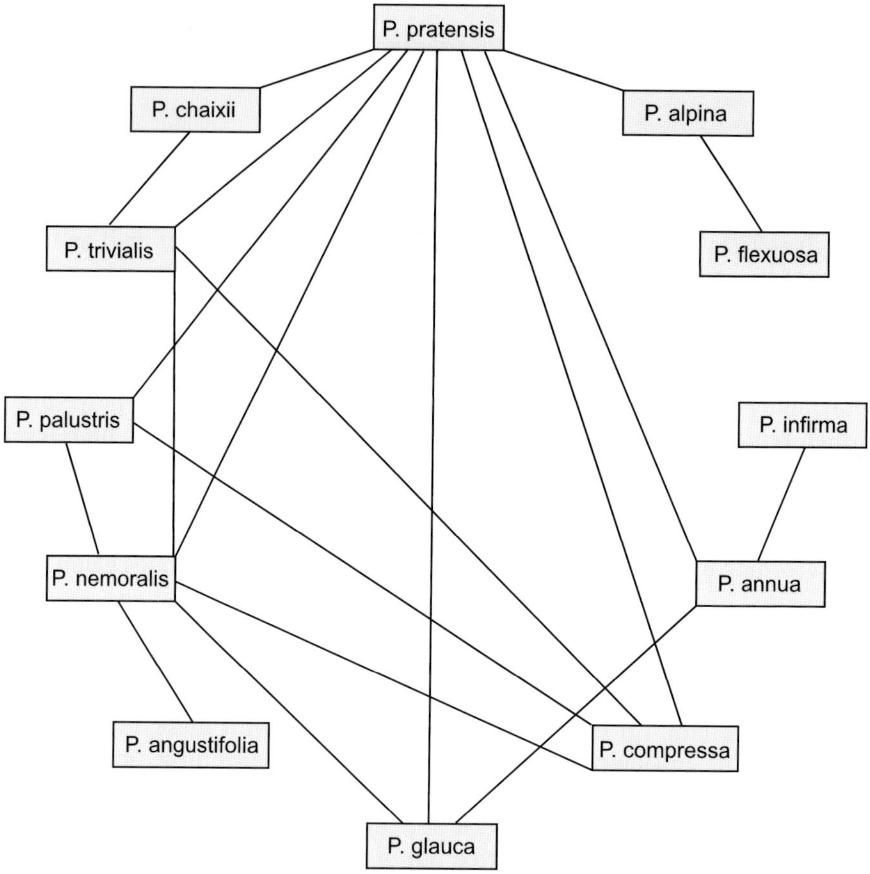

Fig. 2 Wild-to-wild bridges: possible intrageneric hybridization in the grass genus *Poa* [112–160]

Figure 3 gives some further examples of wild-to-wild inter- and intrageneric hybrid complexes in three common European genera of *Poaceae*. The reproductive compatibility and hybrid viability (even at intrageneric level) between *Lolium, Festuca* and *Vulpia* are very well documented (e.g. [51, 76–78]). Intrageneric spontaneous hybrids between *Festuca* and *Lolium* (= x *Festulolium*) are not rare (see also Table 3), they can be fertile and have an ability to backcross with either of the parents [71, 79]. The commonest x *Festulolium* in Europe is the hybrid between *F. pratensis* x *L. perenne* (= x *Festulolium loliaceum*) which can be found in different types of pastures and meadows from Norway to Italy [51, 80, 81]. Figure 3 shows additionally that there are certain species complexes where hybrid combinations are possible in all directions. This is the case for example in the following five species: *Festuca pratensis, F. arundinacea, F.*

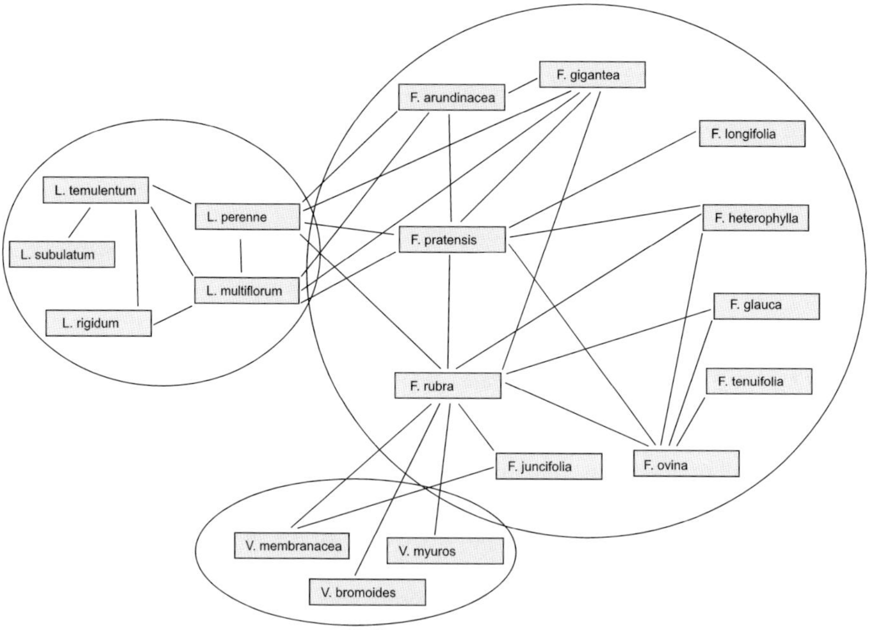

Fig. 3 Wild-to-wild bridges: possible intra- and intergeneric hybridization between selected grass genera: *Lolium*, *Festuca* and *Vulpia* [112–160]

gigantea, *Lolium perenne* and *L. multiflorum*. The close relation of these species could be also demonstrated experimentally [82]. Several authors proposed even to join both genera or to move some *Festuca* species into genus *Lolium* [83, 84]. Additionally, some studies on the chromosome structure of *F. pratensis*, *L. perenne* and *L. multiflorum*, concluded that there are practically no barriers for gene exchange between these species [71, 78].

The majority of species represented in Figs. 2 and 3 fit very well the general characteristic of taxa predisposed for hybridization [12, 52]. They are outcrossing with incomplete reproductive isolation between species, they are mainly perennials with well-developed vegetative spread. Further, they are wind-pollinated and the pollen dispersal up to 21 km has been shown (e.g. for *Agrostis* [16]). Thus, the geographic proximity as well as pollination does not represent any constrains. Additionally, they flower over a very long time period from May till August, thus even at a local scale the phenological overlapping is very common.

The examples described above illustrate clearly that in selected vascular plant families and genera we could potentially expect a stepping-stone spread and exchange of genes with unpredictable effect. Absolute containment of transgenes will be in such taxa practically impossible. Therefore, more experimental and descriptive work has to be done in order to evaluate the existence and importance of wild-to-wild bridges among a spectrum of taxonomic plant groups as broad as possible.

4 Genetic and Ecological Consequences on Wild Relatives

4.1 Genetic and Ecological Consequences of Outcrossing

Outcrossing in plants may have different impacts, depending on the relatedness of the taxa. When a single species is involved, chromosomes are homologous and pair regularly. On the contrary, when related taxa hybridize, recombinations occur between homeologous chromosomes with the possible consequence of irregular pairing, leading to unbalanced gametes with reduced fertility.

Hybridization between crops and wild relatives is a very ancient phenomenon which has been investigated for a long period from an agronomist point of view, as gene flow from the wild species to the crop might lead to reduced yield and loss of the genetic purity of the cultivated varieties. More recently, while GE plants have been developed, agronomists and ecologists have been concerned by the consequences of transgene escape into non-transgenic crop fields or in wild relatives.

Hybridization has genetic and ecological consequences. Genetic consequences may be defined as the effects of the insertion of the genes in the target species itself and on the expression of genes. On the other hand, ecological consequences are considered here as direct or indirect effects on fitness. We discuss below the two types of consequences separately.

4.2 Consequences of the Transgene

The transgene itself may have genetic consequences for the recipient plants by interacting with other genes and leading to untargeted effects. In order to investigate this aspect, among many others, *Arabidopsis thaliana* has been used as a model species. Metzdorff et al. [85] analyzed, using cDNA microarrays, six independently transformed *A. thaliana* lines characterized by modified flavonoid biosynthesis. Although these transgenic lines possessed different types of integration events, no unintended effects were identified.

Genetic transformation could also affect fitness, and may be in this case associated with a physiological cost. For example, significant reduction of fitness was observed repeatedly associated with resistance to herbicide. Bergelson et al. [86] observed for *A. thaliana* a 34% reduction in seed production for a mutant acetolactase synthetase gene that confers resistance to the herbicide chlorsulfuron, in comparison to the non-transgenic lines. This cost in fitness was caused by pleiotropic effects due to the presence of the resistance genes itself, while no cost was associated with the expression of kanamycin resistance. Purrington and Bergelson [87] obtained similar results by comparing mutant and transgenic herbicide resistant lines in two different environmental conditions: with or without fertilizer treatments. The cost of resistance appeared in both treatments for the transgenic line, while no cost was associated with the mutant line in the high fertilizer treatment. Other untargeted effects may appear, such as the change in outcrossing rate observed in an outdoor experiment involving transgenic *A. thaliana*, without the proof that it was caused by the transgene itself [88].

Transgenesis may have also unexpected effects on crops. For example, the lignin content of Bt corn was significantly higher than that of non-Bt corn [89]. A change in lignin content might affect the action of herbivores and have ecological consequences.

Moreover, Prescott et al. [90] demonstrated that post-translational modification of a plant protein (α-amylase inhibitor-1 from the common bean (*Phaseolus vulgaris*)) led to the synthesis of a structurally modified form of the protein in pea (*Pisum sativum*). This protein showed altered antigenic properties. While this example concerns human health, it shows that untargeted effects of transgenic plants on protein expressions occur. Consequently, we can infer that similar effects could lead to changes in ecological properties.

4.3 Consequences for Wild Relatives

Introgression involves chromosome segments containing possibly several genes. Therefore, the consequences of introgression will depend on the genes included in the introgressed segment, and on the site of introgression in the recipient species (linkage to other crop genes, pleiotropy). Similarly, introgression from a transgenic crop to a wild relative will also depend on the insertion site of the transgenic line. Genetic consequences are expected to be those of conventional lines, except for the effect of the transgene itself.

4.3.1 Inference from Natural Observations

Natural hybridization is frequent in nature and the fitness of hybrids may be lower, equal or higher than that of their parents [91]. Hybrid inferiority has been recognized as a rule for a long time. More recently, the importance of hybridization for evolution and speciation has emerged (e.g. [2, 59, 60]). Hybrids may be at the origin to new lineages, which may lead to new species. Speciation might be either progressive or abrupt, when polyploids are formed by chromosome doubling. Because of their different chromosome number, allopolyploids are reproductively isolated, at least partly, from their parents.

According to Burke and Arnold [92], different genetic mechanisms operate behind low and high fitness of the hybrids. Hybrid inferiority would be caused in most cases by negative epistasis, while heterosis would be mostly the consequences of the segregation of additive genetic factors.

4.3.2 Inference from Conventional (Non-Transgenic) Wild x Crop Hybrids

Hybridization followed by repetitive backcrosses lead to introgression, the transfer of a part of the genome of one species to another. Depending on the introgressed genes and on their expression, introgression may lead to new characteristics which could affect the ecological properties of the target species. Experimental data produced a broad spectrum of results on the relative fitness between hybrids and their parents.

The effectiveness of gene flow will depend on the viability and the fertility of hybrids and of subsequent backcrosses. Several studies have involved hybrids between a conventional crop and their wild relatives. For example, Hauser et al. [93, 94] have investigated the fitness of F1 hybrids, as well as F2 hybrids and backcrosses between *Brassica rapa* and oilseed rape (*B. napus*) in experimental crosses. Hybrids were as viable as their parents, produced more pods, but these later contained fewer seeds, with an overall fitness that was intermediary to their parents [93]. The fitness of F2 and backcrosses were on average lower, compared to that of their parents, and varied considerably, including individuals as fit as *B. rapa* [94].

In another study, interspecific F1 hybrids between wild and cultivated radishes (*Raphanus raphanistrum* x *R. sativus*) had a lower fitness than the wild plant. Nevertheless, a field experiment was set up with one half containing F1 wild-crop hybrids and the other half wild. After three years, the dominant white color of the flower of the crop persisted at a frequency ranging from 8% to 22% [95]. A similar study on carrot (*Daucus carota*) showed that hybrids between cultivated and wild carrots were more sensitive to frost than the wild parents, which limited their survival [96].

Weed x crop hybrids between *Sorghum halepense* (Johnsongrass) and *S. bicolor* (cultivated sorghum) did not show any difference in fitness, suggesting that in this case, no barriers to gene flow exist [97]. Contrasting with that, a second generation of hybrids (S1 and BC1) between *Lactuca serriola* (prickly lettuce) and *L. sativa* (cultivated lettuce) germinated and survived better than their wild relative. Seed output of both classes of hybrids was greater with *L. sativa* but no significant difference was found with *L. serriola* [98].

4.3.3
Inference from Transgenic Wild x Crop Hybrids

Table 4 summarizes the results of fitness measurements comparing the hybrids of different generations for five crops, partly derived from Hails and Morley [1]. Non-transgenic hybrids have usually lower, equivalent or intermediary fitness then their parents.

A notable exception is the experiment of Guadagnuolo et al. [11] which demonstrated heterosis of hybrids between glyphosate-tolerant maize (*Zea mays*) and teosinte (*Z. mays* ssp. *mexicana*), when compared to the wild parent. Nevertheless, in the absence of selection pressure with herbicide, no difference was detected between transgenic and non-transgenic hybrids.

Sunflowers have been extensively studied. Burke and Rieseberg [10] investigated transgenic sunflower with an inserted gene of oxalate oxidase (OxOx) conferring enhanced white mold resistance in cultivated sunflower. They backcrossed it with wild sunflower. No cost of the transgene was observed in the absence of the pathogen. When the plants were infected, the transgene decreased the probability of infection, this later having a negative effect on seed output. Moreover, the disease effect varied among locations and no generalization was possible. The authors insisted on the necessity of replicating the experiment over space and time, as well as on the importance of genetic background and of environmental conditions.

Snow et al. [99] demonstrated that, in the field, male sterile wild sunflowers introgressed with a Bt transgene produced more inflorescences than those without the transgene. These advantages were related to a decrease in insect damage. Greenhouse experiments did not reveal any fitness cost of the transgene.

In hybrids with transgenic rapeseed, relative fitness varied considerably, depending on the transgene and on the presence of associated selective pressure by insect herbivores. Transgenic hybrids between wild *Brassica rapa* and rapeseed possessing a Bt transgene performed better than non-transgenic hybrids in the presence of herbivores, while their fitness was lower when herbivores were absent, showing a physiological cost of the transgene [100]. On the contrary, performances were equivalent in hybrids with or without a transgene coding for high laurate content [101, 102]. The same was true for glufosinate tolerance, in the absence of herbicide treatment.

For practical reasons, only one transgenic line was used in most of the cited experiments, instead of using so-called "sister lines", possessing the same transgenic construct but in different insertion sites. It is then delicate to assess the effect of the transgenes themselves on the fitness of hybrids and subsequent backcrosses, because the consequences on fitness may not be due necessarily to the transgene, but could depend on its insertion site. Moreover, environmental conditions and the density of plants may also influence relative fitness [10, 103].

Table 4 Comparisons of fitness of F1 and BCs for some crops; data compiled partly from [1]

		NT (non transgenic), T (transgenic)	Type of transgene	Hybrid generation	Fitness (E=equivalent, H=higher, I=intermediary, L= Lower)	Refs.
Oilseed rape						
Brassica napus	*Brassica rapa*	NT		F1	I	[93]
B. napus	*B. rapa*	NT		F2 and BC1	L	[94]
B. napus	*B. rapa*	T	glufosinate tolerance	F1	female fitness L to H: frequency and density dependent, male fitness L	[103]
B. napus	*B. rapa*	T	Bt	F1	H, in presence of herbivores; L in absence	[100]
B. napus	*B. rapa*	T	Bt	BC2	E (low herbivory)	[105]
B. napus	*B. rapa*	T	high laurate	F1	E	[101]
B. napus	*B. rapa*	T	high laurate	F1	E	[102]
B. napus	*B. rapa*	T	glufosinate tolerance	BC3	E	[106]
B. napus	*B. rapa*	T	glufosinate tolerance	BC1	E	[107]
Sunflowers						
Helianthus annuus	*H. annuus* (wild type)	NT		F1	L	[108]
H. annuus	*H. annuus* (wild type)	T	Bt transgene	BC1	H (natural herbivory)	[99]
H. annuus	*H. annuus* (wild type)	T	white mould resistance	BC3	H to E	[10]

Table 4 (continued)

		NT (non transgenic), T (transgenic)	Type of transgene	Hybrid generation	Fitness (E=equivalent, H=higher, I=intermediary, L= Lower)	Refs.
Sugar beet						
Beta vulgaris ssp. *vulgaris*	*B. vulgaris* ssp. *maritima*	NT		F1, F2	E	[109]
B. vulgaris ssp. *vulgaris*	*B. vulgaris* ssp. *maritima*	T	viral resistance	F1, F2	E	[109]
Squash						
Cucurbita pepo	*Cucurbita pepo* (wild type)	NT		F1, F2 and BC1	L (F1) to E (F2 and BC1)	[110]
C. pepo	*C. pepo* (wild type)	T	viral resistance	F1, BC1 and BC2	L (F1) to H (BC1 and BC2) in case of high disease pressure. L for all in case of low disease pressure	[110]
C. pepo	*C. pepo* (wild type)	T	resistant to two pathogenic viruses	F1	L (survival and seed production)	[111]
Maize						
Zea mays	*Z. mays* ssp. *mexicana*	NT		F1	H then wild species	[11]
Zea mays	*Z. mays* ssp. *mexicana*	T	glyphosate tolerance	F1	H then wild species	[11]

Summarizing, costs of the transgene have been observed in some cases but not in all. On the other hand, it is worth noting that in the mentioned studies, positive effects on fitness of transgenic hybrids has been interpreted as the direct effect of the transgene itself.

5 Conclusion

To date, a considerable amount of data has been gathered on potential and actual gene flow between crops and wild relatives. Such knowledge is extremely useful for the risk assessment associated with GE crops, especially when considering the regional component of the floristic composition. The investigation of the case of Switzerland reveals differences with Europe, which modulate the evaluation of risk for several crops. It reveals also that not only the presence of wild relatives differ geographically, but that their genetic composition may vary and strongly influence gene flow, as was illustrated for alfalfa [40].

Other issues merit further investigation. Indeed, transgenes can be transmitted from crops to the closest wild relatives, but can also migrate further to other species, by successive crosses. A bibliographical survey of Poaceae illustrate that hybridization is widespread in some taxonomical groups.

Genetic and ecological consequences of the transgene in a wild species have also been poorly investigated up to now. The few existing studies show different pictures according to the species and the inserted trait. More investigations are needed to dissociate the importance of the insertion site and of the transgenes themselves. Moreover, the fitness of hybrids may vary according to environmental conditions and these interactions merit evaluation in nature. It is interesting to note that several examples demonstrate that transgene expression give advantages to the wild species in cases of selective pressure, as for the Bt gene in the presence of herbivores. While generalization is difficult, any type of transgene which would influence fitness positively, such as for example the resistance to diseases could confer to the wild species a real ecological advantage.

Finally, given the diversity of the results observed in the various studies on gene flow, and on its consequences, it seems almost impossible to address all the questions experimentally. On the contrary, it is probably necessary to produce enough empirical data, in order to build realistic and reliable predictive models.

Acknowledgements This project was partially funded by the National Centre of Competence in Research (NCCR) Plant Survival, a research program of the Swiss National Science Foundation and by the Swiss Federal Office for the Environment. We especially express gratitude to Dr. Christof Sautter for valuable comments offered on earlier versions of the manuscript.

References

1. Hails RS, Morley K (2005) Trends Ecol Evol 20:245
2. Ellstrand NC, Prentice HC, Hancock JF (1999) Ann Rev Ecol Syst 30:539
3. Halfhill MD, Millwood RJ, Rymer PL, Stewart CN (2002) Environ Biosafety Res 1:19
4. Ellstrand NC (2003) Dangerous Liaisons? When Cultivated Plants Mate with Their Wild Relatives. The John Hopkins University Press, Baltimore
5. Stewart CN, Halfhill MD, Warwick SI (2003) Nat Rev Genet 4:806
6. Wilkinson MJ, Elliott LJ, Allanguillaume J, Shaw MW, Norris C, Welters R, Alexander M, Sweet J, Mason D (2003) Science 302:457
7. den Nijs HCM, Bartsch D, Sweet J (2004) Introgression from Genetically Modified Plants into Wild Relatives. CABI Publishing, Oxon, UK
8. Elliott LJ, Mason DC, Wilkinson MJ, Allainguillaume J, Norris C, Alexander M, Welters R (2004) J Appl Ecol 41:1174
9. Reagon M, Snow AA (2006) Am J Bot 93:127
10. Burke JM, Rieseberg LH (2003) Science 23:1250
11. Guadagnuolo R, Clegg J, Ellstrand NC (2006) Ecol Appl 16:1967
12. Chapman MA, Burke JM (2006) New Phytol 170:429
13. Gepts P (2002) Crop Sci 42:1780
14. Ellstrand NC (2003) Philos Trans R Soc Lond B 358:1163
15. Bannert M (2006) Simulation of transgenic pollen dispersal by use of different grain colour maize. Dissertation no. 16508 Swiss Federal Institute of Technology of Zürich
16. Watrud LS, Lee EH, Fairbrother A, Burdick C, Reichman JR, Bollman M, Storm M, King G, Van de Water PK (2004) Proc Natl Acad Sci USA 101:14533
17. Lavigne C, Klein EK, Vallee P, Pierre J, Godelle B, Renard M (1998) Theor Appl Genet 96:886
18. Damgaard C, Kjellsson G (2005) Agric Ecosyst Environ 108:291
19. Hayter KE, Cresswell JE (2006) J Appl Ecol 43:1196
20. Haygood R, Ives AR, Andow DA (2003) Proc R Soc Lond B Biol Sci 270:1879
21. Ellstrand NC, Elam DR (1993) Annu Rev Ecol Syst 24:217
22. Burgess KS, Morgan M, Deverno L, Husband BC (2005) Mol Ecol 14:3471
23. Whitton J, Wolf DE, Arias DM, Snow AA, Rieseberg LH (1997) Theor Appl Genet 95:33
24. Guéritaine G, Sester M, Eber F, Chevre AM, Darmency H (2002) Mol Ecol 11:1419
25. Rieseberg LH, Whitton J, Gardner K (1999) Genetics 152:713
26. Martinsen GD, Whitham TG, Turek RJ, Keim P (2001) Evolution 55:1325
27. Deblock M, Debrouwer D (1993) Planta 189:218
28. Remington DL, Thornsberry JM, Matsuoka Y, Wilson LM, Whitt SR, Doebley J, Kresovich S, Goodman MM, Buckler ES (2001) Proc Natl Acad Sci USA 98:11479
29. Zemetra RS, Hansen J, Mallory-Smith CA (1998) Weed Sci 46:313
30. Seefeldt SS, Zemetra R, Young FL, Jones SS (1997) Weed Sci 46:632
31. Halfhill MD, Richards HA, Mabon SA, Stewart CN (2001) Theor Appl Genet 103:659
32. Schoenenberger N, Felber F, Savova-Bianchi D, Guadagnuolo R (2005) Theor Appl Genet 111:1338
33. Schoenenberger N, Guadagnuolo R, Savova-Bianchi D, Küpfer P, Felber F (2006) Genetics 174:2061
34. Daniell H, Datta R, Varma S, Gray S, Lee SB (1998) Nat Biotechnol 16:345
35. Smith SE (1989) Plant Breed Rev 6:361
36. Huang CY, Ayliffe MA, Timmis JN (2003) Nature 422:72
37. Keenan RJ, Stemmer WPC (2002) Nat Biotechnol 20:215

38. Gressel J (1999) Trends Biotechnol 17:361
39. Al-Ahmad H, Galili S, Gressel J (2005) Planta 272:372
40. Savova D, Rufener Al Mazyad P, Felber F (1996) Bot Helv 106:197
41. Reichman JR, Watrud LS, Lee EH, Burdick CA, Bollman MA, Storm MJ, King GA, Mallory-Smith C (2006) Mol Ecol 15:4243
42. Savova-Bianchi D, Keller-Senften J, Felber F (2002) Weed Res 42:325
43. Fernandes MIBM, Zanatta ACA, Prestes AM, Caetano VR, Barcellos AL, Angra DC, Pandolci V (2000) Genet Mol Biol 23:1051
44. Romero MD, Montes MJ, Sin E, Lopez-Braña I, Duce A, Martin-Sanchez JA, Andrés MF, Delibes A (1998) Theor Appl Genet 96:1135
45. Riley R, Kimber G (1966) Annu Rep Plant Breed Inst 1964–1965:6
46. Choudhary BR, Joshi P, Singh A (2000) Theor Appl Genet 101:990
47. Vroh BI, Baudoinm JP, Hau B, Mergeai G (1999) Euphytica 106:243
48. Chen J, Staub J, Qian Ch, Jiang J, Luo X, Zhuang F (2003) Theor Appl Genet 106:688
49. Laferriere LT, Helgeson JP, Allen C (1999) Theor Appl Genet 98:1272
50. Mallet J (2005) Trends Ecol Evol 20:229
51. Stace C (ed) (1975) Hybridization and the flora of the British Isles. Academic Press, London
52. Ellstrand NC, Whitkus R, Rieseberg LH (1996) Proc Natl Acad Sci USA 93:5090
53. Hardig TM, Brunsfeld SJ, Fritz RS, Morgan M, Orians CM (2000) Mol Ecol 9:9
54. Dumolin-Lapegue S, Kremer A, Petit RJ (1999) Evolution 53:1406
55. Griffin AR, Burgess IP, Wolf L (1988) Austr J Bot 36:41
56. Rieseberg LH (1997) Ann Rev Ecol Syst 28:359
57. Rieseberg LH, Baird SJE, Gardner KA (2000) Plant Mol Biol 42:205
58. Hegarty MJ, Hiscock SJ (2005) New Phytol 165:411
59. Grant V (1981) Plant speciation. Columbia University Press, New York
60. Arnold ML (1997) Natural hybridization and evolution. Oxford University Press, New York
61. Raven PH (1976) Syst Bot 1:284
62. Whitham TG, Morrow PA, Potts BM (1991) Science 254:779
63. Stebbins GL (1959) Proc Am Philos Soc 103:231
64. Abbott RJ (1992) Trends Ecol Evol 7:401
65. Marchant AD, Arnold ML, Willinson P (1988) Heredity 61:321
66. Mathews S, Tsai RC, Kellogg EA (2000) Am J Bot 87:96
67. Davies JI, Soreng RJ (1993) Am J Bot 80:1444
68. Dahlgren RMT, Cliford HT, Yeo PF (1985) The families of the monocotyledons. Springer, Berlin Heidelberg New York
69. Clark LG, Zhang W, Wendel JF (1995) Syst Bot 20:436
70. Stebbins GL (1985) Ann Missouri Bot Gard 72:824
71. Wipff JK (2002) Gene flow in turf and forage grasses (*Poaceae*), In: Ecological and Agronomic Consequences of Gene Flow from Transgenic Crops to Wild Relatives, Workshop Proceedings. The Ohio State University, Columbus, Ohio, p 115–133
72. http://www.isb.vt.edu/
73. Stebbins GL (1950) Variation and evolution in plants. Columbia University Press, New York
74. Clausen J (1961) Euphytica 10:87
75. Muntzing A (1940) Hereditas 15:219
76. Melderis A (1955) Proc BSBI 1:390
77. Trist PJO (1971) Watsonia 8:311
78. Jahuar PP (1975) Chromosoma 52:103

79. Terell EE (1966) Bot Rev 32:138
80. Gymer PT, Whittington WJ (1973a) New Phytol 72:411
81. Gymer PT, Whittington WJ (1973b) New Phytol 72:862
82. Terell EE (1979) Taxonomy, morphology, and phylogeny, In: Buckner RC, Bush LP (eds) Tall Fescue, Chapter 3. Agronomy Series No. 20. American Society of Agronomy, Crop Science Society of America, Soil Science Society of America, Inc., Madison
83. Jaworski A, Sulinowski S, Nowacki E (1975) Genet Pol 16:271
84. Darbyshire SJ (1993) Novon 3:239
85. Metzdorff SB, Kok EJ, Knuthsen P, Pedersen J (2006) Plant Biol 8:662
86. Bergelson J, Purrington CB, Palm CJ, LopezGutierrez JC (1996) Proc R Soc Lond B Biol Sci 263:1659
87. Purrington CB, Bergelson J (1997) Genetics 145:807
88. Bergelson J, Purrington CB, Wichmann G (1998) Nature 395:25
89. Saxena D, Stotzky G (2001) Am J Bot 88:1794
90. Prescott VE, Campbell PM, Moore A, Mattes J, Rothenberg ME, Foster PS, Higgins TJV, Hogan SP (2005) J Agric Food Chem 53:9023
91. Arnold ML, Hodges SA (1995) Trends Ecol Evol 10:67
92. Burke JM, Arnold ML (2001) Ann Rev Genet 35:31
93. Hauser TP, Shaw RG, Østergård H (1998) Heredity 81:429
94. Hauser TP, Jørgensen RB, Østergård H (1998) Heredity 81:436
95. Snow AA, Uthus KL, Culley TM (2001) Ecol Appl 11:934
96. Hauser TP (2002) Conservation Genet 3:75
97. Arriola PE, Ellstrand NC (1997) Ecol Appl 7:512
98. Hooftmann DAP, Oostermeijer GB, Jacobs MMJ, den Nijs HCM (2005) J Appl Ecol 42:1086
99. Snow AA, Pilson D, Rieseberg LH, Paulsen MJ, Pleskac N, Reagon MR, Wolf DE, Selbo SM (2003) Ecol Appl 13:279
100. Vacher C, Weis AE, Hermann D, Kossler T, Young C, Hochberg ME (2004) Theor Appl Genet 109:806
101. Linder CR, Schmitt J (1995) Ecol Appl 5:1056
102. Linder CR, Taha I, Seiler GJ, Snow AA, Rieseberg LH (1998) Theor Appl Genet 96:339
103. Pertl M, Hauser TP, Damgaard C, Jorgensen RB (2002) Heredity 89:212
104. Moser D, Gygax A, Bäumler B, Wyler N, Palese R (2002) Liste Rouge des fougères et plantes à fleurs menacées de Suisse. OFEV, Bern
105. Mason P (2003) Env Biosafety Res 2:263
106. Snow AA, Andersen B, Jorgensen RB (1999) Mol Ecol 8:605
107. Mikkelsen TR, Andersen B, Jorgensen RB (1996) Nature 380:31
108. Snow AA, Moran-Palma P, Rieseberg LH, Wszelaki A, Seiler GJ (1998) Am J Bot 85:794
109. Pohl-Orf M, Morak C, Wehres U, Saeglitz C, Driessen S, Lehnen M, Hesse P, Mücher T, von Soosten S, Schuphan I, Bartsch D (2000) The environmental impact of gene flow from sugar beet to wild beet – an ecological comparison of transgenic and natural virus tolerance genes. In: Fairbairn C, Scoles G, McHughen A (eds) Proc 6th Int Symp Biosafety of Genetically Modified Organisms, Saskatoon, Canada, p 51–55
110. Fuchs M, Chirco EM, Mcferson JR, Gonsalves D (2004) Environ Biosafety Res 3:17
111. Spencer L, Snow A (2001) Heredity 86:694
112. Akerberg E (1942) Hereditas 28:1
113. Akerberg E, Bingefors S (1953) Hereditas 39:1
114. Bangerter EB (1957) Proc BSBI 2:381

115. Benoit PM (1958) Proc BSBI 3:85
116. Benoit PM (1960) Nature Wales 6:59
117. Clapham AR, Tutin TG, Warburg EF (1952) Flora of the British Isles. Cambridge University Press, Cambridge
118. Corkhill L (1945) J Agr 71:465
119. Dale MR, Ahmed MK, Jelenkovic G, Funk CR (1975) Crop Sci 15:797
120. Dandy JE (1958) List of British vascular plants. BSBI, London
121. Druce GC (1919) Rep BEC 5:314
122. Griffiths D (1950) J Agr Sci 40:19
123. Gymer PT (1971) The nature of hybrids between *Lolium perenne* L. and *Festuca pratensis* Huds., PhD Thesis, University of Nottingham
124. Gymer PT, Whittington WJ (1973a) New Phytol 72:411
125. Gymer PT, Whittington WJ (1973b) New Phytol 72:862
126. Hertzsch W (1938) Züchter 10:261
127. Holmberg OR (1922) Hartmans Handbook i Skandinaviens Flora 1:151
128. Hubbard CE (1968) Grasses, 2nd Ed. Penguin Books, Harmondsworth
129. Hylander N (1953) Nordisk Kärlväxtflora 1:356
130. Jahuar PP (1975) Chromosoma 52:103
131. Jenkin TJ (1931) Welsh Pl Breed, Ser H 12:121
132. Jenkin TJ, Thomas PT (1938) J Bot Lond 76:10
133. Jenkin TJ (1954) J Genet 52:282
134. Jenkin TJ (1955a) J Genet 53:442
135. Jenkin TJ (1955b) J Genet 53:467
136. Jenkin TJ (1955c) J Genet 53:94
137. de Langhe J-E, Delvosalle L, Duvigneaud J, Lambinon J, Berghen CV (1973) Nouvelle Flore de la Belgique, du Grand Duché de Luxembourg, du Nord de la France et des Régions voisines. Edition du Patrimoine du Jardin botanique national de Belgique, Bruxelles
138. Lewis EJ (1966) The production and manipulation of new breeding material in *Lolium-Festuca* In: Proc 10th Int Grassland Congr, Helsinki, p 688
139. Linton EF (1907) J Bot Lond 45:296
140. Malik CP, Thomas PT (1966) Chromosoma 18:1
141. Melderis A (1955) Proc BSBI 1:390
142. Melderis A (1957) Proc BSBI 2:243
143. Melderis A (1965) Proc BSBI 6:172
144. Melderis A (1971) Watsonia 8:299
145. Nannfeldt JA (1937) Bot Notiser 1937:1
146. Nilsson F (1935) Hereditas 20:181
147. Nilsson F (1930) Bot Notiser 1930:161
148. Nygren A (1962) Symb Bot Upsal 17:1
149. Patzke E (1970) Biol 51:255
150. Perring FH, Sell PD (1968) Critical supplement to the atlas of the British flora. Nelson, London
151. Petch CP, Swann EL (1968) Flora of Norfolk. Jarrold and Sons Ltd, Norwich
152. Peto FH (1933) J Genet 28:113
153. Stace CA (1975) Hybridization and the Flora of the British Isles. Academic Press, London
154. Stace CA, Cotton R (1974) Watsonia 10:119
155. Terrell EE (1966) Bot Rev 32:138
156. Trist PJO (1971) Watsonia 8:311

157. Tutin TG (1957) Watsonia 4:1
158. Ullmann W (1936) Herbage Rev 4:105
159. Webster MMCC (1973) Watsonia 9:390
160. Willis AJ (1967) Proc BSBI 6:386

Invited by: Professor Sautter

Adv Biochem Engin/Biotechnol (2007) 107: 207–234
DOI 10.1007/10_2007_047

Published online: 24 March 2007

Assessing Effects of Transgenic Crops on Soil Microbial Communities

Franco Widmer

Molecular Ecology, Agroscope Reckenholz-Tänikon Research Station ART,
Reckenholzstrasse 191, 8046 Zürich, Switzerland
franco.widmer@art.admin.ch

Abstract Deleterious effects of transgenic plants on soils represent an often expressed concern, which has catalyzed numerous studies in the recent past. In this literature review, studies addressing this question have been compiled. A total of 60 studies has been found, and their findings as well as their analytical approaches are summarized. These studies analyzed the effects of seven different types of genetically engineered traits, i.e., herbicide tolerance, insect resistance, virus resistance, proteinase inhibitors, antimicrobial activity, environmental application, and biomolecule production. Sixteen genetically engineered plant species were investigated in these studies including corn, canola, soybean, cotton, potato, tobacco, alfalfa, wheat, rice, tomato, papaya, aubergine, and silver birch. Many of these plants and traits have not been commercialized and represent experimental model systems. Effects on soil microbial characteristics have been described in various studies, indicating the sensitivity and feasibility of the analytical approaches applied. However, classification of the observed effects into acceptable and unacceptable ones has not been possible so far. Establishment of validated indicators for adverse effects represents a scientific challenge for the near future, and will assist risk assessment and regulation of transgenic plants commercially released to the field.

Keywords Literature review · Transgenic plants · Genetically engineered plants · Soil microbial ecology · Effects · Analytical methods · Microbial community profiling

1 Introduction

Along with the increasing potential for widespread commercial use and the potential benefits of transgenic crops, considerable concerns on their safety have been raised including safety aspects relating to their potential impact on the environment [1–5]. Besides ecological effects on organisms in the aboveground compartment [6], effects on the below-ground compartment, in particular on soil and rhizosphere, have gained increasing attention [5, 7–14]. Soil has been recognized as a valuable resource for agriculture and therefore it has to be managed in a sustainable manner in order to maintain its quality [15].

Soil quality can be described based on physical, chemical, and biological soil characteristics, of which soil biological, and in particular soil microbiological, characteristics are least defined [8]. Soil microorganisms, however, play a central role in soil processes, such as nutrient cycling, formation of soil structures, and transformation of pollutants, but they can also act as plant growth promoters or plant pathogens [11]. Soils have been reported to contain up to 10^{10} microbial cells per cm^3, but this number may largely depend on the ecosystem and soil type [16]. Arable soils appear to contain fewer microbial cells and a lower diversity than pasture or forest soils, but still hundreds or thousands of genotypes have been described in agriculturally managed soils. In an agricultural field experiment near Basel in Switzerland [17], for example, a bacterial species richness of about 1300 has been estimated in soils after 25 years of defined organic or conventional management [18]. However, only about 1% of these bacteria can be cultivated with current cultivation techniques [19–21], representing a strong bias for cultivation-based approaches. This has resulted in a large knowledge gap on functions and physiologies of soil microorganisms. Therefore, it is currently difficult or impossible to define which of these organisms are essential for a specific ecosystem, and which changes in abundance and diversity represent damage of a given soil ecosystem [8, 16, 19]. These difficulties represent the main reason for the lack of soil quality definitions that are based on specific microbiological indicators. Recent efforts to define microbial indicator groups of known importance for soil quality resulted in the definition of a set of functional groups [8]. Potential microbial soil indicators perform key soil functions and include mycorrhizal fungi, nitrogen-fixing bacteria, ammonia-oxidizing bacteria, decomposers of recalcitrant organic compounds, and antagonists of plant pathogens as well as plant pathogens. The indicator quality of these groups still remains to be confirmed and possibly other groups like plant growth-promoting bacteria, entomopathogenic fungi, or endophytic microorganisms may also represent indicators for soil quality.

A large body of information has been accumulated over the years, which indicates that agricultural management has various effects on microbial soil characteristics. It has been clearly demonstrated that soil tillage [22–25], fertilization [17, 26–29], and crops [27, 29–31] have strong influences on microbial soil characteristics. In addition, soil type has been shown to be a major determinant of the microbiota present in soils [26, 32–35]. These many factors that influence soil microbial communities, their possible interactions, and the large proportion of unknown members of these communities represent the main current obstacles to the definition of "healthy" soil microbial communities and to defining soil quality based on soil microbiological characteristics. Nevertheless, if a specific factor causes a significant change in soil microbial communities, this may be taken as an indication that microbiological soil characteristics are affected; however, further analyses may be required to assess the importance of such an effect. This approach also represents the basis for assessing the effects of transgenic plants on soil ecosystems.

2 Methods Used for Assessing Soil Microbial Characteristics

A number of techniques have been developed for the analysis of microbiological soil characteristics and to assess the impact environmental or anthropogenic factors may have on soil ecosystems [36–39]. Commonly applied bulk soil microbial parameters include determination of activities of enzymes, such as phosphatases, proteases, cellulases, and dehydrogenases [37, 40], total microbial biomass [41–45], and basal soil respiration [39, 46, 47].

Many cultivation-dependent approaches are available to retrieve microorganisms from soils, but they are all affected by the cultivation bias precluding analysis of the entire microbial diversity in soils [16, 21, 38]. More detailed analyses based on simultaneous cultivation of soil microbial communities on an array of specific substrates have improved the capacity of cultivation-dependent analyses [48–54]. However, these substrate utilization-based approaches, referred to as community-level physiological profiles (CLPP) or community-level substrate utilization (CLSU), are restricted to culturable microorganisms.

During the last two decades, the development of cultivation-independent approaches, which are based on analyses of molecular markers, has allowed for less biased analyses of soil microbial communities [36, 38]. The main targets for this type of analysis are DNA and fatty acids, which can both be directly extracted from soil [55–57] and which contain information on the organisms present.

With the DNA-based approach, specific marker genes can be analyzed in soil DNA extracts by means of polymerase chain reaction (PCR) amplification. Commonly used marker genes are those encoding ribosomal RNA, as these represent phylogenetic markers [20, 58, 59] that allow identification of the microorganisms present in soil. In addition, functional markers are available, which allow the analysis of specific functional microbial groups in soils, such as nitrogen-fixing bacteria [60, 61] and ammonium-oxidizing bacteria [62, 63]. Commonly applied analytical procedures to resolve community structures of detected soil microbial communities are based on genetic profiling, which allows the display of the various microbial genotypes present in a soil. Different genetic profiling approaches for soil microorganisms have been developed, all targeting differences in marker gene DNA sequences. Restriction fragment length polymorphisms (RFLP), amplified ribosomal DNA restriction analysis (ARDRA) [57, 64], or terminal RFLP (T-RFLP) [65, 66] distinguish sequences based on different locations of restriction enzyme recognition sites. Denaturant gradient gel electrophoresis (DGGE) [30, 67, 68] relies on differences in DNA duplex stability, while single-strand conformation polymorphism (SSCP) analysis [69] detects differences in secondary structures of single-stranded DNA. Length heterogeneity (LH) [70] and ribosomal intergenic spacer analysis (RISA) [66, 71, 72] resolve length differences of the amplified marker gene fragment. Such marker gene profiles allow analysis of the presence and abundance of specific genotypes in a PCR amplification product; however, they do not provide quantitative analyses of microbial groups in soil. For this purpose quantitative PCR approaches have been developed. In addition, PCR-based genome-typing protocols have been developed, which allow for the distinction of microbial isolates according to profiles of amplified genomic sequences. Examples of these analyses are ERIC- and BOX-PCR [73].

For the fatty acid-based approach, cell wall lipids of the soil microbiota are extracted from soil. Specific marker fatty acids have been identified for specific groups of microorganisms, which can analytically be distinguished and thereby allow monitoring of their presence or abundance in soil. Similar to genetic profiling, fatty acid profiles can be used to detect differences or changes in microbiological soil characteristics. Often used fatty acid profiling approaches are based on separation and identification of fatty acids, and are referred to as phospholipid fatty acid (PLFA) analysis [52–54, 74–77] or fatty acid methyl ester (FAME) analysis [49, 76, 78–80]. Both the classical cultivation-dependent and the more recent molecular approaches are currently applied to assessing the effects of environmental and anthropogenic factors on microbiological soil characteristics.

3 Assessing Effects of Transgenic Crops on Soil Microbiota

Potential negative effects of transgenic plants on soil microorganisms may arise in different ways and may differ from those of conventional agricultural practices. Some soil microorganisms live in close contact with plants or plant debris in the field and may thereby be exposed to specific active compounds of transgenic plants. Genetically engineered plants may also release their engineered gene products via their root exudates into the soil [81–86], which then may persist in soil and retain their activities [87–90]. In turn, these substances may affect microorganisms in soils even after plants have been removed, and may possibly alter populations of plant-beneficial or plant-pathogenic microorganisms. These different routes of exposure may have different effects on soil microorganisms and can be grouped into four categories.

Effects on soil microorganisms may arise from:

1. Close contact with the living plant (e.g., rhizosphere or plant interior)
2. Close contact with plant litter, also after crops have been harvested
3. Exposure to released transgene products that may persist in soil
4. Horizontal gene transfer from transgenic plants, their debris, or released DNA to soil indigenous microorganisms

In addition to direct effects of the transgene product, altered plant physiologies due to indirect (pleiotropic) effects of genetic transformation may occur and affect plant-associated microorganisms [91–93]. Therefore, it is not sufficient to test for effects of the engineered trait, but rather to assess the performance of engineered plants in the environment and to compare them with different genotypes of conventionally bred cultivars. The following sections will focus on studies assessing effects of transgenic plants and transgene products on soil microbial community structures and performances. Persistences of transgenes [94–100] and transgene products [89, 90, 101–106] as well as horizontal gene transfer [96, 97, 107–116] have been summarized by others and are not within the scope of this review of the scientific literature on the effects of transgenic plants on soil microbial communities, although some of the studies presented have also addressed one or more of these aspects.

Altogether, these considerations reveal that a variety of potentially negative impacts of transgenic plants on soil microbial communities have been suggested. As a result, an increasing number of studies have been published in the recent past, and much information has been collected by applying an array of different analytical methods and approaches. In the following, an overview of results published in the scientific literature will be provided. This review does not claim to be complete, but it may be representative of the type of research questions addressed to assess potential effects of different transgenic traits in various plant types on soil microbial communities and the type of results one may expect from these analyses.

4 Studies Assessing Effects of Transgenic Plants on Soil Microbial Community Structures

4.1 Crops Engineered for Herbicide Tolerance

In 2005, the worldwide area planted with herbicide tolerant transgenic crops was 73.8 million hectares [117]. Commercialized genetically engineered herbicide tolerant traits confer resistance to active substances, such as glufosinate (e.g., Basta®) and glyphosate (e.g., Roundup®). Effects of genetically engineered herbicide tolerant plants on soil microbial ecology have been addressed in numerous studies and the data have been reviewed by others [9, 13]. Nine studies addressing the effects of genetically engineered herbicide tolerant plants on soil microbiota have been identified in the scientific literature. Five of them were performed with canola (*Brassica* spp.), three with corn (*Zea mays*), and one with soybean (*Glycine max*), all of them addressing potential effects of the engineered trait on soil microbial communities.

4.1.1
Herbicide Tolerant Canola

In a field study, endophytic and rhizosphere microbial communities of different canola cultivars were analyzed, including Quest, a transgenic variety tolerant to the herbicide glyphosate [49]. CLSU and FAME profiling were used to characterize the microbial communities associated with the root interior and the rhizosphere. Both techniques revealed differences between both the endophytic and the rhizosphere microbial communities of the transgenic cultivar Quest and nontransgenic cultivars. In a follow-up study, the differences between endophytic bacteria were confirmed with FAME analysis of root endophytic bacterial isolates [78]. These results were supported by a 2-year field study with four genetically modified and four conventional canola varieties [118], where CLSU and FAME analyses of microbial community structures in the roots and the rhizospheres were applied. These analyses revealed that the root interior and rhizosphere bacterial community associated with the genetically modified varieties differed from those of conventional varieties. Finally, in a contained experiment with genetically modified glufosinate tolerant canola and associated herbicide applications [119, 120], shifts in rhizosphere bacterial communities and *Pseudomonas* population structures were assessed. Rhizosphere soil was sampled at different stages of plant development and DGGE analyses of PCR-amplified 16S rRNA gene fragments were applied. Bacterial community and *Pseudomonas* profiling revealed slightly altered microbial communities in the rhizosphere of transgenic plants; however, effects were minor when compared to the plant developmental stage-dependent shifts. In addition, invertase, urease, and alkaline phosphatase activities were significantly enhanced in the rhizospheres of senescent transgenic plants when compared to wild-type plants. The authors attributed the observed differences between transgenic and wild-type lines to altered root exudation of the herbicide tolerant canola.

4.1.2
Herbicide Tolerant Corn

Bacterial communities in rhizospheres of field-grown glufosinate tolerant transgenic corn have been assessed with SSCP analyses of PCR-amplified bacterial 16S rRNA gene fragments [121]. Neither the genetic modification nor the use of glufosinate affected the rhizosphere bacterial SSCP profiles. On the other hand, clear differences have been detected between the rhizospheres of corn and sugar beet controls, clearly demonstrating the sensitivity of the approach chosen. A less pronounced but significant difference has been detected at certain growth stages in rhizosphere samples obtained from the fine root fraction. The same authors confirmed this lack of strong effects of herbicide tolerant corn on rhizosphere bacterial communities, again based on bacterial SSCP profiles [122]. In this second study no differences between cultivars or treatments have been detected that were greater than the variability between replicates. These data have been confirmed with results obtained in glasshouse and field studies using CLSU and DGGE analyses [123]. These analyses have revealed stronger differences in bacterial community structures among soil textures than among corn genotypes, leading the authors to the conclusion that bacterial communities in corn rhizospheres may be more strongly affected by soil texture than by the engineered herbicide tolerant trait.

4.1.3
Herbicide Tolerant Soybean

Bradyrhizobium japonicum contains a glyphosate-sensitive 5-enolpyruvylshikimic acid-3-phosphate synthase and it has been demonstrated that at high concentrations, glyphosate may result in growth inhibition or death of *B. japonicum*. In an overview, the effects of glyphosate application and glyphosate resistant soybean on its nitrogen-fixing symbiont were assessed [13]. In glasshouse studies it has been shown that nitrogenase activity in glyphosate tolerant soybean could be transiently inhibited after glyphosate application, indicating the potential for reduced nitrogen fixation in the herbicide tolerant soybean system.

4.2
Crops Expressing Insecticidal Bt Toxins

In 2005, the worldwide area planted with insect resistant transgenic crops was 26.3 million hectares [117]. Commercialized genetically engineered insect resistant crops expressed variations of insecticidal proteins from subspecies of *Bacillus thuringiensis*, i.e., Bt toxins. There are numerous studies on potential effects of Bt crops on soil microbial communities and several reviews have summarized their results [5–9, 11, 89]. In the present review, 21 studies addressing effects of Bt-based genetically engineered insect resistant plants on soil microbiota were identified in the scientific literature. Four of them assessed the effects of purified Bt toxins, while 13 investigated insect resistant corn (*Z. mays*) and two each investigated insect resistant cotton (*Gossypium* spp.) or rice (*Oryza* spp.).

4.2.1
Purified Bt Toxin

In a study on effects of transgenic cotton lines on microbiological soil characteristics, control treatments included application of purified toxin from *B. thuringiensis* subsp. *kurstaki* (Btk toxin) [124]. In this study, the purified

Btk toxin did not reveal any significant effects on plate counts of bacteria and fungi. These data were confirmed in a study where purified Bt toxin was added to soil [125], revealing no significant changes in the numbers of culturable bacteria in rhizosphere soil, except for nitrogen-fixing bacteria at a Bt toxin concentration of 500 ng/g soil. This represents a rather high Bt toxin concentration when compared to the concentrations detected in the rhizospheres of Bt corn, which have been reported to reach 10 ng/g soil as detected with an enzyme-linked immunosorbent assay (ELISA) [90]. Furthermore, the effects of purified Bt toxins have been assessed by in vitro studies on a selection of microorganisms [126]. A variety of bacteria, fungi, and algae were tested in pure and mixed cultures, as well as in disk-diffusion and sporulation assays with purified free and clay-bound Bt toxins. In these analyses no antibiotic effects were detected. Recently, however, preliminary results on the significantly higher half-life of the herbicides glyphosate and glufosinate in soils amended with purified Btk toxin have been reported [127]. These authors concluded that the absence of Btk toxin effects on soil microbial biomass and the rapid decrease of insecticidal activity as determined with a bioassay may indicate indirect effects of the Btk toxin on soil properties and/or mechanisms that influence herbicide degradation.

4.2.2
Insect Resistant Bt Corn

Extractable lipids in Bt and conventional corn shoots and soil were analyzed at harvest [128]. Concentrations of total alkenes, *n*-alkanes, and *n*-fatty acids were increased in soils planted with Bt corn, while unsaturated fatty acid contents were higher in soil planted with non-Bt corn. Cumulative CO_2 released from soils was lower under Bt corn, indicating that cultivation of Bt corn may reduce microbial activity. In a growth chamber experiment [52], two lines of Bt corn expressing either Cry1Ab or Cry1F were compared with nontransgenic isolines in three soil types. PLFA profiles of bulk soil and CLPP profiles of rhizosphere soils revealed only for one soil significant Bt corn effects in the rhizosphere. Expression of Bt toxin also significantly reduced the presence of eukaryotic PLFA biomarkers in bulk soils; however, it remained unclear which eukaryotes they represented. From this data, the authors concluded that potential effects of Bt corn on soil and rhizosphere microbial communities may be small. In a glasshouse study [129], the effects of Bt-176 corn on the rhizosphere bacterial community have been analyzed. Bacterial plate counts and CLSU revealed no significant differences between plant genotypes. On the other hand, differences between the rhizosphere and bulk soil bacterial communities could be detected. Bacterial RISA revealed differences in the rhizosphere communities at different plant growth stages, as well as between Bt-176 and control corn. The authors attributed the different bacterial communities in the rhizospheres to altered root exudates of the transgenic corn.

In soil samples from field trials with Bt corn expressing Cry1Ab [54], microbial communities were analyzed based on CLSU and PLFA profiling, as well as based on protozoa analyses. Two occasions were reported when soil protozoa populations under Bt corn were reduced as compared to non-Bt corn. CLSU profiling revealed one occurrence of differences between Bt and control corn cultivars. The effects of Bt corn were classified by the authors as small and comparable to the variation expected in these agricultural systems. Finally, PLFA profiling was used to analyze the microbial communities in soil samples collected from fields with Bt corn [77]. Analyses revealed a reduction in fungal abundance and ratios of Gram-positive to Gram-negative bacteria in soils from Bt corn; however, the authors have stated that the causes of these observed effects remained unknown and require more detailed investigations.

Several studies focused on mycorrhizal fungi for the assessment of effects of Bt corn. An experimental model system was used to study the effects of root exudates of Bt corn on different stages of the life cycle of the arbuscular mycorrhizal fungal species *Glomus mosseae* [130]. Root exudates of Bt-176 corn significantly affected presymbiotic hyphal growth and development of appressoria, as compared to Bt-11 and control corn. Differential hyphal morphogenesis occurred irrespective of Bt or control corn, suggesting that Bt toxin did not interfere with fungal host recognition mechanisms. In microcosm experiments [131], the impact of genetically modified Bt-11 and Bt-176 corn on soil respiration, rhizosphere, and bulk soil bacterial communities, and the mycorrhizal symbiont *G. mosseae*, were further assessed. DGGE profiling of bacterial 16S rRNA gene fragments showed differences in rhizosphere bacterial communities associated with all corn lines, while mycorrhizal colonization was significantly reduced for Bt-176 corn only. Additional glasshouse experiments confirmed the differences between Bt and non-Bt corn, and addition of Bt corn residues to soil affected soil respiration, bacterial communities, and mycorrhizal establishment. In another study [132], colonization with arbuscular mycorrhizal fungi and activity of rhizosphere soil microbiota were determined during growth of Cry1Ab-expressing Bt corn in the field. The results suggested that Bt corn and conventional corn may differ in their C/N ratios. In addition, reduced colonization with arbuscular mycorrhizal fungi and increased microbial activity were found during early Bt corn development. The authors conclude that genetic transformation might have led to changes in plant physiology and root-exudate composition, which in turn may have affected symbiotic and rhizosphere microorganisms.

There are also a number of studies which revealed no effects of Bt corn on microbiological soil characteristics. For instance, soils were planted with Cry1Ab-expressing Bt corn or amended with Bt corn biomass and compared to controls [84]. Analysis was based on a cultivation-dependent approach and revealed no significant differences in the plate counts for bacteria and fungi, as well as in the numbers of protozoa between rhizosphere soil of Bt and control corn. Also, amendment with plant biomass of these plants revealed no different effects. The authors concluded that the Bt protein in corn-root exudates and plant biomass appeared not to be toxic to protozoa, bacteria, and fungi. In a field study [133], effects of corn rootworm (*Diabrotica* spp.) resistant Bt corn expressing Cry3Bb and application of the insecticide tefluthrin were assessed. Analyses included soil microbial biomass, N-mineralization potential, short-term nitrification rate, basal respiration, and bacterial community structures based on T-RFLP analysis. The data showed no effects of Bt corn on microbial measures or bacterial community structures when compared to the near isoline. T-RFLP analysis revealed substantial temporal differences and tefluthrin application reduced soil respiration. The authors concluded that *Diabrotica* resistant Bt corn may pose little or no threat to soil microbiology. In other field studies with Cry1Ab-expressing Bt corn [90], the persistence of Bt toxin in soil and the effects on rhizosphere bacterial communities were assessed. An improved ELISA method for Bt toxin quantification and SSCP analysis of PCR-amplified bacterial 16S rRNA gene fragments were used. Despite the presence of Cry1Ab protein in the rhizosphere of Bt corn, effects on bacterial community structures were small when compared to other factors, such as plant age or field heterogeneities. In glasshouse and field studies [123], bacterial diversity in Bt and conventional corn rhizospheres was determined. CLSU profiling and DGGE of PCR-amplified 16S rRNA gene fragments allowed differentiation of bacterial communities among different soil textures but not among corn varieties. From these results the authors concluded that cultivation of transgenic varieties may not affect rhizosphere bacterial communities. These results have been supported by a recent glasshouse experiment [134] on the effects of Bt corn (Cry1Ab) and the insecticide deltamethrin on soil microbiota. The Bt trait induced an increase of protozoa, but significant effects on soil microbial community structure, as determined with CLSU and PLFA analyses, were caused only by soil type and plant growth stages. Results from this glasshouse experiment were in broad agreement with those of a field experiment using the same plant material grown in the same soils [54].

4.2.3
Insect Resistant Bt Cotton

Leaves of Bt cotton and purified Bt toxin were placed in soil and analyses included plate counts of indigenous soil bacteria and fungi [124]. Two transgenic Btk cotton lines caused at some sampling dates a transient increase in total bacterial and fungal plate counts. Transient changes in bacterial species composition, measured by biochemical tests, CLSU, and ARDRA, were also observed for the two transgenic Bt cotton lines. In contrast, neither a third Btk cotton line nor the purified Btk toxin had any significant effects. The plant line specificity of the effects observed and the lack of effects of purified Bt toxin suggested that the observed effects may have resulted from pleiotropism. In another field study [125], the effects of Bt cotton on soil microbiota were monitored. Bt toxin released from cotton roots was determined with ELISA and microbial populations were analyzed by selective plate counts. Significant differences were found in the culturable fraction of bacteria in the rhizosphere of Bt cotton, but no significant differences were found after the growing season. Furthermore, only the addition of 500 ng purified Bt toxin per gram of soil resulted in significant changes in the numbers of culturable nitrogen-fixing bacteria. From these results the authors concluded that pleiotropic factors might possibly be involved.

4.2.4
Insect Resistant Bt Rice

In a laboratory study, the impacts of Bt rice straw amendment on biological activities in water-flooded soil were investigated [135]. The results revealed some differences in protease, neutral phosphatase, and cellulase activities between soil amended with Bt rice straw or conventional rice straw but none of these differences was persistent. However, differences in dehydrogenase activity, methanogenesis, hydrogen production, and anaerobic respiration persisted over the course of the experiment. The results indicated shifts in microbial populations or changes in their metabolic abilities. In a second study [136], nonpersisting occasional differences in plate counts of actinomycetes, fungi, anaerobic fermentative bacteria, denitrifying bacteria, hydrogen-producing acetogenic bacteria, and methanogenic bacteria were detected between the paddy soils amended with Bt rice straw and conventional rice straw. These effects supported the results from the first study.

4.3
Crops Engineered for Virus Resistance

Genetically engineered virus resistances still play a minor role in commercialized crops but may have an increasing potential in the future. Commercially cultivated transgenic virus resistant crops were papaya (*Carica papaya*) and squash (*Cucurbita pepo*) [117]. Three studies investigated the effects of plants engineered for virus resistance on microbiological soil characteristics, i.e., two using the papaya system and one focusing on potato (*Solanum tuberosum*).

4.3.1
Virus Resistant Papaya

The influence of papaya ringspot virus resistant transgenic papaya on soil microorganisms was assessed in soil samples collected from areas where transgenic and a nontransgenic papaya were grown for 9 years, as well as from an area where no papaya was grown [137]. Moisture content, pH value, total organic carbon contents, and total nitrogen contents were comparable among the soils. Plate counts for fungi and actinomycetes were highest in upper-layer soils around transgenic papaya plants and lowest in lower-layer soils where no papaya was grown. ARDRA, T-RFLP, and DGGE analyses revealed that soil bacterial communities shared more than 80% similarity between the areas planted with transgenic and nontransgenic papaya. The authors concluded that cultivation of virus resistant transgenic papaya had only limited effects on soil microorganisms. In another study [138], soil was amended with replicase-transgenic or nontransgenic papaya under field conditions and soil properties, microbial communities, and enzyme activities were recorded. Total nitrogen in soils planted with transgenic papaya was significantly different. Significant increases in plate counts of bacteria, kanamycin resistant bacteria, actinomycetes, and fungi were found in the transgenic papaya treatment. Transgenic papaya and nontransgenic papaya induced significantly different activities for arylsulfatase, polyphenol oxidase, invertase, cellulase, and phosphodiesterase. The authors concluded that transgenic papaya could alter soil chemical properties as well as microbial communities.

4.3.2 Virus Resistant Potato

PLFA profiling was used to analyze the effects of potato virus Y (PVY) resistant transgenic potato on microbial communities in field soils [77]. A decrease of fungal abundance in soils from PVY resistant transgenic potato was reported. Contrasting differences were found in the ratios of Gram-positive to Gram-negative bacteria in different depth layers. The authors state that the causes of these differences are unclear and require further investigation.

4.4 Crops Engineered with Proteinase Inhibitors

4.4.1 Proteinase Inhibitor-Expressing Potato

Only one study has been found in the scientific literature that investigated effects of crops expressing protein inhibitors on soil microbial communities [74]. The system studied was a potato line genetically engineered for nematode resistance with chicken egg white cystatin (two lines) or modified rice cystatin (one line), both being cysteine proteinase inhibitors. In a field study, the effects of these plants were compared to those of aldicarb, an oxime carbamate nematicide. PLFA analyses were used to investigate effects on soil bacteria and fungi. In the first year, chicken egg white cystatin-expressing potato was tested, and one transgenic line revealed increases and the other decreases in fungal marker fatty acid abundance later in the growing season. In the second year, rice cystatin-expressing potato and nematicide treatment were used. The nematicide treatment reduced the bacterial fraction of the microbial community, while the rice cystatin-expressing potato reduced both bacterial and fungal community components. No differences in the rate of leaf litter decomposition were observed. These results indicated that nematicide use and different transgenic lines may differentially influence components of soil microbial communities without affecting soil functions such as litter decomposition.

4.5 Crops Engineered with Antimicrobial Activities

Some microorganisms represent a serious threat to agriculture and genetic engineering offers an attractive approach for the production of disease resistant crops. A variety of genes coding for antimicrobial proteins from plant, animal, or microbial origin have been used to transform crops for improved disease resistance. However, justified concerns have been raised that these engineered traits may also affect beneficial microorganisms, such as mycorrhizae, rhizobia, plant growth promoting microorganisms, or other microorganisms improving plant health, as well as microorganisms involved in plant litter decomposition and nutrient cycling [139]. Genetically engineered antimicrobial activities in crops actually have the potential for direct effects on endophytic, epiphytic, symbiotic, rhizosphere, and soil microorganisms. An increasing number of studies that have assessed potential effects of antimicrobial transgenic plants have become available, and for this review 17 were found in the scientific literature. Ten studies assessed potato (*Solanum tuberosum*), two each aubergine (*Solanum melongena*) and woodland tobacco (*Nicotiana sylvestris*), and one each tomato (*Solanum lycopersicum*), wheat (*Triticum aestivum*), and silver birch (*Betula pendula*).

4.5.1 T4 Lysozyme-Expressing Potato

Potato genetically engineered to produce bacteriophage-derived T4 lysozyme for enhanced bacterial resistance has gained considerable attention in the scientific literature, with seven studies on potential effects of this genetically engineered trait in potato.

Changes in plant-associated bacterial populations were monitored during a 2-year field release of T4 lysozyme potato [140]. No significant differences in aerobic plate counts were observed between transgenic and control plant lines. In addition, no significant differences in counts of auxin-producing and phytopathogen-antagonistic isolates were found. Among 28 different antagonistic species isolated, seven were found only on control plants. However, the observed difference was minor when compared to the variability during the monitoring period. In further field evaluations, T4 lysozyme tolerant mutants of two antagonistic plant-associated bacterial isolates, i.e., *Pseudomonas putida* and *Serratia grimesii*, were used for seed tuber inoculation of transgenic T4 lysozyme-expressing and control potato [141]. Both introduced isolates colonized the rhizo- and geocaulosphere of transgenic and control potato. At flowering, significantly higher plate counts of the T4 lysozyme tolerant *P. putida* were recovered from transgenic T4 lysozyme-expressing potato. Effects of the inoculants on the indigenous bacterial community were monitored by DGGE analysis of PCR-amplified fragments of 16S rRNA gene fragments. Neither dominance of the inoculated strains nor differences between inoculated and uninoculated potato were detected. In order to further assess potential effects of these plants, a total of 68 representative bacterial strains of the group enterics and pseudomonads were isolated from parental and transgenic T4 lysozyme-expressing potato [79]. They were identified with FAME analysis and typed by phenotypic profiling, i.e., antagonistic activity, auxin production, and sensitivity to T4 lysozyme, as well as genotypic profiling with BOX-PCR. The majority of identified bacterial groups included isolates from all potato lines analyzed. The authors concluded that no correlations between bacterial types and plant genotype have evolved. In a further study [142], a polyphasic approach was chosen to analyze rhizosphere bacterial communities of the T4 lysozyme-expressing potato lines and control plants at two field sites over 3 years. The polyphasic approach included heterotrophic plate counts, identification of isolates with FAME, CLSU profiling, DGGE profiling of bacteria, actinomycetes, and alpha- and beta-proteobacteria, as well as DNA sequence analyses. These analyses revealed that environmental factors related to season, field site, or year of sampling influenced the rhizosphere communities but no effects related to the T4 lysozyme trait were detected. Some transgenic line-specific differences were attributed by the authors to pleiotropic effects of genetic engineering.

In a separate series of glasshouse experiments, the effects of T4 lysozyme-expressing potato on rhizosphere bacterial communities in different soil types were analyzed [143]. Soil enzyme activities involved in C-, P-, and N-nutrient cycles and T-RFLP-based bacterial community structures were assessed. Transgenic potato induced differences in soil enzyme activities and structures of rhizosphere bacterial communities; however, the impact of genetic modification was only transient, minor, or comparable to those caused by soil type, control plant genotype, vegetation stage, and pathogen exposure. In a second study [144], the authors assessed the effects of T4 lysozyme expression on endophytic bacteria of potato by using T-RFLP profiling and 16S rRNA gene sequencing approaches. Genetic transformation induced small differences in the endophytic community structures; however, the effects were also minor or comparable to the variations induced by environmental factors. The authors pointed out that effect assessment studies on transgenic crops should include different environmental factors in order to allow for ranking of potential transgene-related effects.

Finally, a detailed study has been performed in order to assess potentially harmful effects of T4 lysozyme exudation into the rhizosphere [145]. The bactericidal effects of T4 lysozyme-expressing transgenic potato were assessed in a model system with *Bacillus subtilis* associated with hair roots of the plants. Significantly decreased survival of *B. subtilis* was observed on hair roots of T4 lysozyme-expressing potato lines in this model system. However, the authors conclude that no strong negative effects of T4 lysozyme-producing potato on soil bacteria may be expected in the field.

4.5.2
Attacin/Cecropin-Expressing Potato

Attacin and cecropin are insect-derived proteins with antimicrobial activity [146]. In glasshouse experiments the effects of attacin/cecropin-expressing potato on rhizosphere bacterial communities were analyzed [143]. Soil enzyme activities of C-, P-, and N-nutrient cycles, as well as bacterial community structures based on T-RFLP profiling of bacterial 16S rRNA genes, were determined. In general, the T4 lysozyme trait used in the same experiment (described in the previous section) had stronger effects than the attacin/cecropin trait. Therefore, for the attacin/cecropin trait the authors conclude that the effects of genetic modification were not stronger than those of soil type, control plant genotype, vegetation stage, and pathogen exposure. In the follow-up study [144], effects of attacin/cecropin-expressing transgenic potato on endophytic bacteria were assessed by using T-RFLP profiling. Similar to the T4 lysozyme-expressing potato used in parallel (described in the previous section), attacin/cecropin expression induced differences in the community structures of endophytic bacteria; however, also in this case the effects were not larger than those of environmental factors.

4.5.3
Magainin II-Expressing Potato

The gene for the antimicrobial peptide magainin II, derived from the African clawed toad (*Xenopus laevis*), showed in vitro activity against a range of microorganisms including rhizosphere isolates [147]. Transgenic potato expressing magainin II revealed increased resistance to the bacterial potato pathogen *Erwinia carotovora*. Bacterial and fungal plate counts on different media were used to assess effects on communities associated with magainin II-producing potato plants. Analyses revealed no significant differences in the bacterial counts from leaf and root samples. Higher numbers of culturable fungi were detected in root samples and significantly lower numbers of total bacteria in tubers of magainin II-expressing transgenic potato.

4.5.4
Defensin-Expressing Aubergine

Dm-AMP1 is an antifungal plant defensin from *Dahlia merckii* [148]. Aubergine transformed for constitutive expression of defensin showed increased resistance to the pathogenic fungus *Botrytis cinerea* [86]. The protein was released in root exudates of the transformed plants and was active in reducing the growth of the pathogenic fungus *Verticillium alboatrum*, whereas it did not interfere with recognition responses and symbiosis establishment by the arbuscular mycorrhizal fungus *Glomus mosseae*. In an experimental model system, effects of Dm-AMP1 defensin-containing root exudates of aubergine on different stages of the life cycle of *G. mosseae* were assessed [130]. In contrast to root exudates of Bt-176 corn (described above), no differences were found in mycelial growth and fungal host recognition mechanisms.

4.5.5
gox Gene-Expressing Tomato

It has been shown that plants engineered with the *gox* gene, encoding for the enzyme D-glucose oxygen 1-oxidoreductase, have elevated H_2O_2 concentrations and exhibit increased resistance to plant pathogens [149]. Also, transgenic tomato plants engineered with the *gox* gene revealed increased resistance to some pathogens and contained more nitrogen and insoluble lignin as well as less soluble protein than control plants [93]. Soil amended with leaves from the transgenic tomato line revealed reduced soil respiration during the first 2 days of incubation. This was explained by the authors as due to the different composition of the plant material, and may in part be related to the pleiotropic effects of genetic engineering.

4.5.6
KP4-Expressing Wheat

The viral *kp4* gene is derived from a double-stranded RNA virus infecting corn smut (*Ustilago maydis*). The gene codes for a "killer protein" (KP) that is expressed in infected *U. maydis* and inhibits growth of competing *U. maydis* strains lacking viral infection [150, 151]. Further studies revealed that KP4 may reversibly block ion channels and therefore represents a growth inhibitor rather than a killing protein [152–154]. Various bacteria and fungi have been tested for their sensitivity toward KP4 and it has been shown that only specific genera of the order Ustilaginales, which cause smut and bunt diseases in cereals, were affected [155, 156]. In order to test whether KP4 may confer resistance to specific fungal diseases if used in genetic plant engineering, spring wheat varieties were transformed with the *kp4* gene [157]. Tests in climate chambers using artificial infection with stinking smut (*Tilletia caries*) revealed an increased resistance of the *kp4*-transgenic wheat [157, 158]. These results have been confirmed in a recent study, where two *kp4*-transgenic spring wheat varieties were grown in a convertible glasshouse, allowing exposure of the plants to the open environment but also more strictly containing them if required [156, 159]. Laboratory bioassays and glasshouse studies with the collembola *Folsomia candida* revealed no effects of the *kp4*-transgenic wheat, while differences among different wheat varieties were detected [160]. In addition, investigations in the convertible glasshouse system [159] revealed no effects of the *kp4* transgene on wheat infesting insects, i.e., aphids and cereal leaf beetle. T-RFLP profiling of PCR-amplified ribosomal RNA gene fragments from bacteria (PCR primers 27F and 1378R) and from fungi (PCR primers NS1 and FR1) was performed on DNA extracted from bulk soil (Widmer, unpublished results). Statistical analysis revealed a significant ($p < 0.05$) effect of the factors "sampling time" and "wheat variety" on community structures of bacteria and fungi. For the factors "kp4-transgenic" and "*Tillecia tritici* inoculation", no significant effects on bacterial and fungal community structures were detected.

4.5.7 Chitinase-Expressing Woodland Tobacco

Transgenic woodland tobacco (*Nicotiana sylvestris*) expressing different tobacco chitinases was used to assess colonization of the root system by the root pathogenic fungus *Rhizoctonia solani* and the vesicular–arbuscular mycorrhizal symbiont *Glomus mosseae* [161]. Transgenic *N. sylvestris* expressing the vacuolar tobacco chitinase A or an N-terminally truncated version of this chitinase revealed increased resistance to *R. solani*. Transgenic *N. sylvestris* expressing a C-terminally truncated chitinase A showed no enhanced resistance. All these transgenic *N. sylvestris* lines were equally well colonized by *G. mosseae*, indicating that expression of the different chitinase A forms did not interfere with the vesicular–arbuscular mycorrhizal symbiosis. In a second study, genetically engineered tobacco lines expressing various pathogenesis-related proteins (PRs) were examined [162]. Constitutive expression of various tobacco PRs, e.g., a basic tobacco chitinase, a cucumber acidic chitinase, and some combinations of these genes, did not affect the time course or the final level of colonization by *G. mosseae*.

4.5.8 Glucanase-Expressing Woodland Tobacco

In the same series of experiments as described in the previous section, *N. sylvestris* lines genetically transformed to express glucanases were examined [162]. Constitutive expression of various tobacco PRs, a cucumber acidic chitinase, or a basic beta-1,3-glucanase had no effects on the time course or the final level of colonization by vesicular–arbuscular mycorrhizal symbiont *G. mosseae*. Only constitutive expression of the acidic isoform of tobacco PR-2, a protein with beta-1,3-glucanase activity, resulted in delayed colonization by *G. mosseae*.

4.5.9
Chitinase-Expressing Silver Birch

The decomposition of leaf litter from eight transgenic silver birch lines expressing sugar beet chitinase IV was studied in a field experiment [46]. Leaf litter decomposition was analyzed based on total litter mass, as well as content of total microbial biomass (based on substrate-induced respiration (SIR)), of total fungal biomass (based on ergosterol contents), and microbial activity (basal respiration). Mass loss of transgenic leaf litter did not differ from controls and no differences in either the fungal or total microbial biomass were recorded. Only one transgenic birch line, which revealed high levels of chitinase IV expression, showed distinct temporal dynamics of nematode populations and might indirectly indicate microbial differences in this litter. This transgenic line-specific effect may be related to the high level of chitinase expression, but may also be an indication of possible pleiotropic effects of genetic engineering on plant litter quality.

4.6
Plants Engineered for Environmental Applications

Genetic engineering allows the introduction of traits into plants of interest for specific applications in the environment. Examples of these applications are phytoremediation or the specific design of plant rhizospheres. Five studies in the scientific literature addressed potential impacts of such plants on soil microbiology. Two studies each used birdsfoot trefoil (*Lotus corniculatus*) and tobacco (*Nicotiana tabacum*), while another was based on black nightshade (*Solanum nigrum*).

4.6.1
Opine-Expressing Birdsfoot Trefoil and Black Nightshade

Opines are small amino acid and sugar conjugates representing specific bacterial growth substrates. Therefore, transgenic plants releasing these substances in root exudates may support and select specific natural or possibly recombinant rhizobacteria. This principle was verified with opine-producing transgenic birdsfoot trefoil (*L. corniculatus*), by showing that engineered opine-producing plants induced targeted alterations in their root-associated bacterial communities, resulting in a stimulation of opine-utilizing populations [81]. The fate of the opine-utilizing bacterial community was investigated over time and under different experimental conditions [82]. After removal of the transgenic plants the density of opine-utilizing bacteria in the fallow soils remained unchanged. If soils were replanted with wild-type *L. corniculatus*, only specific bacterial populations able to utilize opines were affected. Numbers of nopaline utilizers decreased to the level of control plants, while numbers of mannopine utilizers remained at an intermediate level. Data indicated that opine-utilizing bacterial populations respond to engineered plant exudation and that certain alterations in bacterial communities may be more responsive to crop rotation. In a follow-up study [85], it was demonstrated that this targeted alteration of rhizosphere bacterial populations was not restricted to the *L. corniculatus* system described above, but was also effective with black nightshade (*S. nigrum*) growing in another soil type.

4.6.2
Ferritin-Expressing Tobacco

Ferritin is a ubiquitous iron storage protein that plays an important role in iron metabolism. Its ability to sequester iron provides a dual function, i.e., iron detoxification and iron storage [163]. Ferritin over-expressing transgenic tobacco with activated iron transport and increased iron phytoextraction may deplete iron from its rhizosphere and thus select for iron stress-sensitive rhizobacteria [164]. Plate counts on media depleted in or supplemented with iron were used to determine the abundance of iron stress-sensitive bacteria. Ferritin over-expressing tobacco revealed highest iron accumulation at the floral bud stage, the time point when the density of iron stress-sensitive bacteria recovered was significantly increased in the rhizosphere of these plants. This effect, however, was soil type and plant stage dependent. Data indicate that the ferritin over-expressing transgenic tobacco plants are able to extract and accumulate more iron from the rhizosphere, and that they may select in their rhizosphere for bacteria that are less susceptible to iron stress.

4.6.3
Metallothionein-Expressing Tobacco

Transgenic tobacco plants expressing yeast metallothionein in combination with a polyhistidine cluster displayed increased accumulation of and tolerance for cadmium. These plants were assessed for their effects on the arbuscular mycorrhizal fungus *Glomus intraradices* in a pot experiment [165]. Mycorrhiza tended to decrease the phytoextraction efficiency of the transgenic tobacco, while it increased that of nontransgenic plants at cadmium levels in the soil that are inhibitory to growth of tobacco. These results indicate that plant–mycorrhiza interactions may be important for phytoextraction efficiencies and may depend on the plant genotype.

4.7 Crops Engineered for the Production of Biomolecules

Plants may be genetically engineered for the production of substances of interest to industry. These substances may include enzymes for industrial processes and also substances for pharmaceutical applications. In addition, plants may also be engineered for improved nutritional quality. Reports on the assessment of such plants concerning their effects on the soil ecosystem are still scarce in the scientific literature. Four studies were retrieved and three analyzed effects of transgenic alfalfa (*Medicago sativa*), while another study assessed transgenic potato (*S. tuberosum*).

4.7.1 Alpha-Amylase- or Lignin Peroxidase-Expressing Alfalfa

Rhizosphere bacterial communities of two transgenic alfalfa lines, one expressing bacterial alpha-amylase and the other expressing fungal lignin peroxidase, were analyzed based on CLSU profiling [50]. Genetic profiles of bacterial consortia present in individual substrate wells of the Biolog GN plates were determined with ERIC-PCR. Analyses of the CLSU profiles indicated consistent differentiation of lignin peroxidase-expressing alfalfa plant rhizospheres. ERIC-PCR profiles revealed consistent differences in the substrate-specific consortia enriched from each alfalfa genotype rhizosphere. ERIC-PCR-based typing of bacterial isolates obtained from substrate wells suggested that a limited number of bacteria were responsible for specific substrate utilization. Data suggested that transgenic plant genotypes may affect rhizosphere microorganisms. The same transgenic alfalfa plants were then used in a field study in combination with recombinant *Sinorhizobium meliloti* in order to assess the effects of genetically engineered organisms on soil ecosystems [51]. Analyses included plant shoot weight, soil chemistry, enzyme activities, SIR, plate counts of indigenous soil bacteria and fungi, and counts of protozoa, as well as CLSU and ARDRA profiling of soil bacterial communities. The lignin peroxidase-expressing plants had significantly lower shoot weight, and higher nitrogen and phosphorus contents. Significantly higher soil pH and lower activity of soil dehydrogenase and alkaline phosphatase were associated with the lignin peroxidase-expressing alfalfa, while plate counts for culturable aerobic spore-forming and cellulose-utilizing bacteria were increased. CLSU profiles distinguished all three alfalfa genotypes, but particularly the lignin peroxidase-expressing plants. Counts for protozoa, ARDRA profiles of indigenous soil bacteria, and SIR rates were not significantly affected by any of the transgenic alfalfa treatments. The primary effects observed were associated with the transgenic lignin peroxidase-expressing alfalfa and could possibly be explained by the different plant characteristics found for this genotype.

4.7.2
Ovalbumin-Expressing Alfalfa

Fourteen genetically modified lines of alfalfa, expressing a methionine-rich ovalbumin from Japanese quail, were evaluated for nodulation ability and plate counts for different aerobic bacteria in the rhizosphere [166]. Higher counts of ammonifying, spore-forming, denitrifying, and nitrifying bacteria were observed in the rhizospheres of transgenic lines, while counts for cellulolytic bacteria and *Azotobacter* spp. were decreased. In spite of some differences in colony numbers in samples isolated from the rhizosphere of transgenic and nontransgenic alfalfa plants, no statistically significant difference between individual lines could be detected.

4.7.3
Potato Engineered for Altered Starch Composition

For certain industrial products amylopectin offers advantages as compared to amylose. Therefore a transgenic potato line was developed, which expresses the mRNA for the granule-bound starch synthase gene (*gbss*) also in an antisense direction. This approach results in reduced levels of this enzyme in the potato plants and in a modified starch composition. This transgenic potato line was evaluated for its effects on soil microbiology [167] by analyzing DGGE profiles of bacterial and fungal ribosomal RNA gene fragments from bulk soil samples. It was shown that no significant differences between the two cultivars and the transgenic line were found. For rhizosphere samples only bacterial DGGE patterns differentiated the conventional cultivar *SOLANA* from those of the parental line *SIBU* and the transgenic line *SIBU S1*, and the sequence of the differentiating band showed the highest similarity with *Enterobacter amnigenus*. *Pseudomonas*-specific DGGE analyses revealed differences between the rhizospheres of the transgenic line *SIBU S1* and the parental cultivar *SIBU*. However, this analysis also indicated clear differences between the cultivars *SOLANA* and *SIBU*. Therefor, differentiation detected for the transgenic line was comparable to the one observed among different cultivars.

5
Conclusions

A large number of studies on effects of genetically engineered plants on soil microbiological characteristics have become available in the recent past. Many different types of engineered plant species and traits have been studied, and a large array of classical and more recently developed tools have been applied. Effects have been studied in laboratory systems, microcosms, glasshouse systems, and in the field. In many studies differences in soil microbiological characteristics between soils planted with transgenic or control plants have been detected, although a large number of studies found no effects. In some studies effects detected were compared to those caused by environmental factors or other crop types, and often it was found that these factors have a greater influence on soil microbiological characteristics than the genetically engineered trait. Effects are often restricted to the rhizosphere of the transgenic plants or to the time period when these plants were present. In addition, many of the effects described were based on analyses of symbiotic microorganisms that live in close association with the plant, such as mycorrhizal fungi or endophytic bacteria. Effects on these plant-associated microorganisms may well be disadvantageous for the crop itself, and may therefore represent a potential economic restriction for applying this crop rather than a concern for the ecosystem. In addition, many of the effects found appeared spatially and temporally limited, and therefore may also potentially affect the transgenic crop itself. However, these conclusions represent hypotheses based on the available data, and need scientific evaluation in future experiments and monitoring of fields planted with genetically engineered crops. The result of the studies presented here indicate that the tools for sensitive detection of changes in soil microbiological characteristics are available; however, they also reveal that at present it is very difficult or impossible to define which alterations in these characteristics may represent unacceptable damage to a soil system. This limitation becomes evident from the scientific literature presented here, as no study reported damage of a soil system, but rather potentially adverse effects. The definition and identification of indicators that quantitatively represent soil quality or soil damage will be one of the great scientific challenges in soil ecology for the near future. Analyses of soil microbial communities with their diverse functions may allow for the identification of such indicators, which may be used in specific diagnostics for assessing damage to soil systems.

Acknowledgements The author acknowledges support from Martin Hartmann and Kaspar Schwarzenbach for searching the scientific literature. Jürg Enkerli, Olivier Sanvido, and Christof Sautter are acknowledged for critically reading and commenting on this manuscript. Partial funding for this study was received from the Swiss Federal Office for the Environment (FOEN).

References

1. Dale PJ, Clarke B, Fontes EMG (2002) Nat Biotechnol 20:567
2. Conner AJ, Glare TR, Nap J-P (2003) Plant J 33:19
3. Sayre P, Seidler RJ (2005) Plant Soil 275:77
4. Snow AA, Andow DA, Gepts P, Hallerman EM, Power A, Tiedje JM, Wolfenbarger LL (2005) Ecol Appl 15:377

5. Velkov VV, Medvinsky AB, Sokolov MS, Marchenko AI (2005) J Biosci 30:515
6. O'Callaghan M, Glare TR, Burgess EPJ, Malone LA (2005) Ann Rev Entomol 50:271
7. Bruinsma M, Kowalchuk GA, van Veen JA (2003) Biol Fertil Soils 37:329
8. Kowalchuk GA, Bruinsma M, van Veen JA (2003) Trends Ecol Evol 18:403
9. Dunfield KE, Germida JJ (2004) J Environ Qual 33:806
10. Lynch JM, Benedetti A, Insam H, Nuti MP, Smalla K, Torsvik V, Nannipieri P (2004) Biol Fertil Soils 40:363
11. Motavalli PP, Kremer RJ, Fang M, Means NE (2004) J Environ Qual 33:816
12. Sessitsch A, Smalla K, Kandeler E, Gerzabek MH (2004) In: Gillings M, Holmes A (eds) Plant microbiology. BIOS, Oxford, p 55
13. Zablotowicz RM, Reddy KN (2004) J Environ Qual 33:825
14. Giovannetti M, Sbrana C, Turrini A (2005) Riv Biol Biol Forum 98:393
15. Francaviglia R (ed) (2004) Agricultural impacts on soil erosion and soil biodiversity: developing indicators for policy analysis. OECD, Paris, p 654
16. Torsvik V, Ovreas L, Thingstad TF (2002) Science 296:1064
17. Mäder P, Fliessbach A, Dubois D, Gunst L, Fried P, Niggli U (2002) Science 296:1694
18. Hartmann M, Widmer F (2006) Appl Environ Microbiol 72:7804
19. Amann R (2000) Syst Appl Microbiol 23:1
20. Amann R, Ludwig W (2000) FEMS Microbiol Rev 24:555
21. Colwell RR, Grimes DJ (2000) Nonculturable microorganisms in the environment. ASM, Washington DC
22. Lupwayi NZ, Rice WA, Clayton GW (1998) Soil Biol Biochem 30:1733
23. Frey SD, Elliott ET, Paustian K (1999) Soil Biol Biochem 31:573
24. Drijber RA, Doran JW, Parkhurst AM, Lyon DJ (2000) Soil Biol Biochem 32:1419
25. Carpenter-Boggs L, Stahl PD, Lindstrom MJ, Schumacher TE (2003) Soil Till Res 71:15
26. Bossio DA, Scow KM, Gunapala N, Graham KJ (1998) Microbial Ecol 36:1
27. Marschner P, Crowley D, Yang CH (2004) Plant Soil 261:199
28. Hartmann M, Fliessbach A, Oberholzer HR, Widmer F (2006) FEMS Microbiol Ecol 57:378
29. Widmer F, Rasche F, Hartmann M, Fliessbach A (2006) Appl Soil Ecol 33:294
30. Pesaro M, Widmer F (2006) Appl Environ Microbiol 72:37
31. Smalla K, Wieland G, Buchner A, Zock A, Parzy J, Kaiser S, Roskot N, Heuer H, Berg G (2001) Appl Environ Microbiol 67:4742
32. Marschner P, Yang CH, Lieberei R, Crowley DE (2001) Soil Biol Biochem 33:1437
33. Schutter ME, Sandeno JM, Dick RP (2001) Biol Fertil Soils 34:397
34. Buyer JS, Roberts DP, Russek-Cohen E (2002) Can J Microbiol 48:955
35. Girvan MS, Bullimore J, Pretty JN, Osborn AM, Ball AS (2003) Appl Environ Microbiol 69:1800
36. Hill GT, Mitkowski NA, Aldrich-Wolfe L, Emele LR, Jurkonie DD, Ficke A, Maldonado-Ramirez S, Lynch ST, Nelson EB (2000) Appl Soil Ecol 15:25
37. Emmerling C, Schloter M, Hartmann A, Kandeler E (2002) J Plant Nutr Soil Sci 165:408
38. Kirk JL, Beaudette LA, Hart M, Moutoglis P, Khironomos JN, Lee H, Trevors JT (2004) J Microbiol Methods 58:169
39. Joergensen RG, Emmerling C (2006) J Plant Nutr Soil Sci 169:295
40. Caldwell BA (2005) Pedobiologia 49:637
41. Anderson JPE, Domsch KH (1978) Soil Biol Biochem 10:215
42. Vance ED, Brookes PC, Jenkinson DS (1987) Soil Biol Biochem 19:703
43. Bailey VL, Peacock AD, Smith JL, Bolton H (2002) Soil Biol Biochem 34:1385

44. Leckie SE, Prescott CE, Grayston SJ, Neufeld JD, Mohn WW (2004) Soil Biol Biochem 36:529
45. Dilly O (2006) Biol Fertil Soils 42:241
46. Vauramo S, Pasonen HL, Pappinen A, Setala H (2006) Appl Soil Ecol 32:338
47. Visser S, Addison JA, Holmes SB (1994) Can J Forest Res 24:462
48. Garland JL, Mills AL (1991) Appl Environ Microbiol 57:2351
49. Siciliano SD, Theoret CM, de Freitas JR, Hucl PJ, Germida JJ (1998) Can J Microbiol 44:844
50. Di Giovanni GD, Watrud LS, Seidler RJ, Widmer F (1999) Microbial Ecol 37:129
51. Donegan KK, Seidler RJ, Doyle JD, Porteous LA, Digiovanni G, Widmer F, Watrud LS (1999) J Appl Ecol 36:920
52. Blackwood CB, Buyer JS (2004) J Environ Qual 33:832
53. Soderberg KH, Probanza A, Jumpponen A, Baath E (2004) Appl Soil Ecol 25:135
54. Griffiths BS, Caul S, Thompson J, Birch ANE, Scrimgeour C, Andersen MN, Cortet J, Messean A, Sausse C, Lacroix B, Krogh PH (2005) Plant Soil 275:135
55. Zelles L (1999) Biol Fertil Soils 29:111
56. Bürgmann H, Pesaro M, Widmer F, Zeyer J (2001) J Microbiol Methods 45:7
57. Widmer F, Fliessbach A, Laczko E, Schulze-Aurich J, Zeyer J (2001) Soil Biol Biochem 33:1029
58. Amann RI, Ludwig W, Schleifer KH (1995) Microbiol Rev 59:143
59. Pace NR (1997) Science 276:734
60. Widmer F, Shaffer BT, Porteous LA, Seidler RJ (1999) Appl Environ Microbiol 65:374
61. Bürgmann H, Widmer F, Von Sigler W, Zeyer J (2004) Appl Environ Microbiol 70:240
62. Sinigalliano CD, Kuhn DN, Jones RD (1995) Appl Environ Microbiol 61:2702
63. Yeager CM, Northup DE, Grow CC, Barns SM, Kuske CR (2005) Appl Environ Microbiol 71:2713
64. Massol-Deya AA, Odelson DA, Hickey RF, Tiedje JM (1995) In: Akkermans ADL, Van Elsas JD, De Bruijn FJ (eds) Molecular microbial ecology manual. Kluwer, Dordrecht, p 1
65. Liu WT, Marsh TL, Cheng H, Forney LJ (1997) Appl Environ Microbiol 63:4516
66. Hartmann M, Frey B, Kolliker R, Widmer F (2005) J Microbiol Methods 61:349
67. Muyzer G, Dewaal EC, Uitterlinden AG (1993) Appl Environ Microbiol 59:695
68. Muyzer G, Smalla K (1998) Antonie Van Leeuwenhoek J Gen Microbiol 73:127
69. Schwieger F, Tebbe CC (1998) Appl Environ Microbiol 64:4870
70. Suzuki M, Rappe MS, Giovannoni SJ (1998) Appl Environ Microbiol 64:4522
71. Fisher MM, Triplett EW (1999) Appl Environ Microbiol 65:4630
72. Ranjard L, Poly F, Lata JC, Mougel C, Thioulouse J, Nazaret S (2001) Appl Environ Microbiol 67:4479
73. Louws FJ, Fulbright DW, Stephens CT, de Bruijn FJ (1994) Appl Environ Microbiol 60:2286
74. Cowgill SE, Bardgett RD, Kiezebrink DT, Atkinson HJ (2002) J Appl Ecol 39:915
75. Baath E, Anderson TH (2003) Soil Biol Biochem 35:955
76. Drenovsky RE, Elliott GN, Graham KJ, Scow KM (2004) Soil Biol Biochem 36:1793
77. Xue K, Luo HF, Qi HY, Zhang HX (2005) J Environ Sci 17:130
78. Siciliano SD, Germida JJ (1999) FEMS Microbiol Ecol 29:263
79. Lottmann J, Berg G (2001) Microbiol Res 156:75
80. Dunfield KE, Germida JJ (2003) Appl Environ Microbiol 69:7310
81. Oger P, Petit A, Dessaux Y (1997) Nat Biotechnol 15:369
82. Oger P, Mansouri H, Dessaux Y (2000) Mol Ecol 9:881
83. Saxena D, Stotzky G (2000) FEMS Microbiol Ecol 33:35

84. Saxena D, Stotzky G (2001) Soil Biol Biochem 33:1225
85. Mansouri H, Petit A, Oger P, Dessaux Y (2002) Appl Environ Microbiol 68:2562
86. Turrini A, Sbrana C, Pitto L, Ruffini Castiglione M, Giorgetti L, Briganti R, Bracci T, Evangelista M, Nuti MP, Giovannetti M (2004) New Phytol 163:393
87. Tapp H, Stotzky G (1998) Soil Biol Biochem 30:471
88. Crecchio C, Stotzky G (1998) Soil Biol Biochem 30:463
89. Stotzky G (2004) Plant Soil 266:77
90. Baumgarte S, Tebbe CC (2005) Mol Ecol 14:2539
91. Saxena D, Stotzky G (2001) Am J Bot 88:1704
92. Bettini P, Michelotti S, Bindi D, Giannini R, Capuana M, Buiatti M (2003) Theor Appl Genet 107:831
93. Marinari S, Messina A, Caccia R, Grego S, Hopkins DW (2004) Commun Soil Sci Plant Anal 35:1851
94. Widmer F, Seidler RJ, Watrud LS (1996) Mol Ecol 5:603
95. Widmer F, Seidler RJ, Donegan KK, Reed GL (1997) Mol Ecol 6:1
96. Paget E, Lebrun M, Freyssinet G, Simonet P (1998) Eur J Soil Biol 34:81
97. Gebhard F, Smalla K (1999) FEMS Microbiol Ecol 28:261
98. Ceccherini MT, Pote J, Kay E, Van VT, Marechal J, Pietramellara G, Nannipieri P, Vogel TM, Simonet P (2003) Appl Environ Microbiol 69:673
99. Crecchio C, Ruggiero P, Curci M, Colombo C, Palumbo G, Stotzky G (2005) Soil Sci Soc Am J 69:834
100. Gulden RH, Lerat S, Hart MM, Powell JR, Trevors JT, Pauls KP, Klironomos JN, Swanton CJ (2005) J Agric Food Chem 53:5858
101. Palm CJ, Donegan K, Harris D, Seidler RJ (1994) Mol Ecol 3:145
102. Head G, Surber JB, Watson JA, Martin JW, Duan JJ (2002) Environ Entomol 31:30
103. Herman RA, Wolt JD, Halliday WR (2002) J Agric Food Chem 50:7076
104. Zwahlen C, Hilbeck A, Gugerli P, Nentwig W (2003) Mol Ecol 12:765
105. Hopkins DW, Gregorich EG (2005) Can J Soil Sci 85:19
106. Wang HY, Ye QF, Wang W, Wu LC, Wu WX (2006) Environ Pollut 143:449
107. Schlüter K, Futterer J, Potrykus I (1995) Biotechnology 13:1094
108. Nielsen KM, Bones AM, Smalla K, Van Elsas JD (1998) FEMS Microbiol Rev 22:79
109. de Vries J, Wackernagel W (1998) Mol Gen Genet 257:606
110. Clerc S, Simonet P (1998) Antonie Van Leeuwenhoek J Gen Microbiol 73:15
111. Lee GH, Stotzky G (1999) Soil Biol Biochem 31:1499
112. Kay E, Vogel TM, Bertolla F, Nalin R, Simonet P (2002) Appl Environ Microbiol 68:3345
113. Meier P, Wackernagel W (2003) Transgenic Res 12:293
114. de Vries J, Wackernagel W (2004) Plant Soil 266:91
115. Heinemann JA, Traavik T (2004) Nat Biotechnol 22:1105
116. Nielsen KM, Townsend JP (2004) Nat Biotechnol 22:1110
117. James C (2005) ISAAA Briefs 34:12
118. Dunfield KE, Germida JJ (2001) FEMS Microbiol Ecol 38:1
119. Gyamfi S, Pfeifer U, Stierschneider M, Sessitsch A (2002) FEMS Microbiol Ecol 41:181
120. Sessitsch A, Gyamfi S, Tscherko D, Gerzabek MH, Kandeler E (2004) Plant Soil 266:105
121. Schmalenberger A, Tebbe CC (2002) FEMS Microbiol Ecol 40:29
122. Schmalenberger A, Tebbe CC (2003) Mol Ecol 12:251
123. Fang M, Kremer RJ, Motavalli PP, Davis G (2005) Appl Environ Microbiol 71:4132
124. Donegan KK, Palm CJ, Fieland VJ, Porteous LA, Ganio LM, Schaller DL, Bucao LQ, Seidler RJ (1995) Appl Soil Ecol 2:111

125. Rui YK, Yi GX, Zhao J, Wang BM, Li ZH, Zhai ZX, He ZP, Li QX (2005) World J Microbiol Biotechnol 21:1279
126. Koskella J, Stotzky G (2002) Can J Microbiol 48:262
127. Accinelli C, Screpanti C, Vicari A, Catizone P (2004) Agric Ecosyst Environ 103:497
128. Dinel H, Schnitzer M, Saharinen M, Meloche F, Pare T, Dumontet S, Lemee L, Ambles A (2003) J Environ Sci Health B 38:211
129. Brusetti L, Francia P, Bertolini C, Pagliuca A, Borin S, Sorlini C, Abruzzese A, Sacchi G, Viti C, Giovannetti L, Giuntini E, Bazzicalupo M, Daffonchio D (2004) Plant Soil 266:11
130. Turrini A, Sbrana C, Nuti MP, Pietrangeli BM, Giovannetti M (2004) Plant Soil 266:69
131. Castaldini M, Turrini A, Sbrana C, Benedetti A, Marchionni M, Mocali S, Fabiani A, Landi S, Santomassimo F, Pietrangeli B, Nuti MP, Miclaus N, Giovannetti M (2005) Appl Environ Microbiol 71:6719
132. Villanyi I, Fuzy A, Biro B (2006) Cereal Res Commun 34:105
133. Devare MH, Jones CM, Thies JE (2004) J Environ Qual 33:837
134. Griffiths BS, Caul S, Thompson J, Birch ANE, Scrimgeour C, Cortet J, Foggo A, Hackett CA, Krogh PH (2006) J Environ Qual 35:734
135. Wu WX, Ye QF, Min H (2004) Eur J Soil Biol 40:15
136. Wu WX, Ye QF, Min H, Duan XJ, Jin WM (2004) Soil Biol Biochem 36:289
137. Hsieh YT, Pan TM (2006) J Agric Food Chem 54:130
138. Wei XD, Zou HL, Chu LM, Liao B, Ye CM, Lan CY (2006) J Environ Sci 18:734
139. Glandorf DCM, Bakker P, VanLoon LC (1997) Acta Bot Neerl 46:85
140. Lottmann J, Heuer H, Smalla K, Berg G (1999) FEMS Microbiol Ecol 29:365
141. Lottmann J, Heuer H, de Vries J, Mahn A, During K, Wackernagel W, Smalla K, Berg G (2000) FEMS Microbiol Ecol 33:41
142. Heuer H, Kroppenstedt RM, Lottmann J, Berg G, Smalla K (2002) Appl Environ Microbiol 68:1325
143. Rasche F, Hodl V, Poll C, Kandeler E, Gerzabek MH, van Elsas JD, Sessitsch A (2006) FEMS Microbiol Ecol 56:219
144. Rasche F, Velvis H, Zachow C, Berg G, Van Elsas JD, Sessitsch A (2006) J Appl Ecol 43:555
145. Ahrenholtz I, Harms K, de Vries J, Wackernagel W (2000) Appl Environ Microbiol 66:1862
146. Taeil K, Young-Joon K (2005) J Biochem Mol Biol 38:121
147. O'Callaghan M, Gerard EM, Waipara NW, Young SD, Glare TR, Barrell PJ, Conner AJ (2004) Plant Soil 266:47
148. Thomma BPHJ, Cammune BPA, Thevissen K (2002) Planta 216:193
149. Wu G, Shortt BJ, Lawrence EB, Levine EB, Fitzsimmons KC, Shah DM (1995) Plant Cell 7:1357
150. Hankin L, Puhalla JE (1971) Phytopathology 61:50
151. Koltin Y (1986) In: Buck KS (ed) Fungal virology. CRC, Boca Raton
152. Bouillet SE, Schoknecht JD, Bozarth RF (1983) Phytopathology 73:787
153. Gu F, Khimani A, Rane SG, Flurkey WH, Bozarth RF, Smith TJ (1995) Structure 3:805
154. Gage MJ, Rane SG, Hockerman GH, Smith TJ (2002) Mol Pharmacol 61:936
155. Koltin Y, Day PR (1975) Appl Microbiol 30:694
156. Schlaich T, Urbaniak BM, Malgras N, Ehler E, Birrer C, Meier L, Sautter C (2006) Plant Biotechnol J 4:63
157. Clausen M, Krauter R, Schachermayr G, Potrykus I, Sautter C (2000) Nat Biotechnol 18:446
158. Sautter C, Kräuter R, Schachermeyr G (2000) Agrarforschung 7:545

159. Romeis J, Waldburger M, Bigler F (2006) IOBC/WPRS Bull 29:129
160. Romeis J, Battini M, Bigler F (2003) Pedobiologia 47:141
161. Vierheilig H, Alt M, Neuhaus JM, Boller T, Wiemken A (1993) Mol Plant Microbe Interact 6:261
162. Vierheilig H, Alt M, Lange J, Gutrella M, Wiemken A, Boller T (1995) Appl Environ Microbiol 61:3031
163. Harrison PM, Arosio P (1996) Biochim Biophys Acta 1275:161
164. Robin A, Vansuyt G, Corberand T, Briat JF, Lemanceau P (2006) Plant Soil 283:73
165. Janouskova M, Pavlikova D, Macek T, Vosatka M (2005) Plant Soil 272:29
166. Faragova N, Farago J, Drabekova J (2005) Folia Microbiol 50:509
167. Milling A, Smalla K, Maidl FX, Schloter M, Munch JC (2004) Plant Soil 266:23

Invited by: Professor Sautter

Adv Biochem Engin/Biotechnol (2007) 107: 235–278
DOI 10.1007/10_2007_048

Published online: 31 March 2007

Ecological Impacts of Genetically Modified Crops: Ten Years of Field Research and Commercial Cultivation

Olivier Sanvido (✉) · Jörg Romeis · Franz Bigler

Agroscope Reckenholz-Tänikon Research Station ART, Reckenholzstr. 191, 8046 Zurich, Switzerland
olivier.sanvido@art.admin.ch

Abstract The worldwide commercial cultivation of genetically modified (GM) crops has raised concerns about potential adverse effects on the environment resulting from the use of these crops. Consequently, the risks of GM crops for the environment, and especially for biodiversity, have been extensively assessed before and during their commercial cultivation. Substantial scientific data on the environmental effects of the currently commercialized GM crops are available today. We have reviewed this scientific knowledge derived from the past 10 years of worldwide experimental field research and commercial cultivation. The review focuses on the currently commercially available GM crops that could be relevant for agriculture in Western and Central Europe (i.e., maize, oilseed rape, and soybean), and on the two main GM traits that are currently commercialized, herbicide tolerance (HT) and insect resistance (IR). The sources of information included peer-reviewed scientific journals, scientific books, reports from regions with extensive GM crop cultivation, as well as reports from international governmental organizations. The data available so far provide no scientific evidence that the cultivation of the presently commercialized GM crops has caused environmental harm. Nevertheless, a number of issues related to the interpretation of scientific data on effects of GM crops on the environment are debated controversially. The present review highlights these scientific debates and discusses the effects of GM crop cultivation on the environment considering the impacts caused by cultivation practices of modern agricultural systems.

Keywords Transgenic crops · Environmental effects · *Bt*-maize · Insect resistance · Herbicide tolerance

Abbreviations

Bt	*Bacillus thuringiensis*
EPA	United States Environmental Protection Agency
FSE	Farm Scale Evaluations
GMO	Genetically modified organism
GM	Genetically modified
GMHT	Genetically modified herbicide tolerant
HT	Herbicide tolerance, herbicide tolerant
IR	Insect resistance, insect resistant
OSR	Oilseed rape
USDA	United States Department of Agriculture

1 Introduction

1.1 GM Crops, Modern Agriculture, and the Environment

The worldwide commercial cultivation of genetically modified (GM) crops has raised concerns about potential adverse effects on the environment from the use of these crops [1–5]. Consequently, the risks of GM crops for the environment, and especially for biodiversity, have been extensively assessed before and during their commercial cultivation. Substantial scientific data on environmental effects of the currently commercialized GM crops are available. Independent from the use of GM crops, modern agricultural systems have considerable negative impacts on global biodiversity [6–11]. On a global scale, the most direct negative impact is due to the considerable loss of natural habitats, which is caused by the conversion of natural ecosystems into agricultural land [9, 12]. The negative impact of modern agricultural systems in Europe cannot be ascribed to only one factor, but is caused by the interaction of a multitude of factors. Several changes in the management of agricultural land over the last century have resulted in a decline in the diversity of plant, invertebrate, and bird species within agro-ecosystems. The significant decline in floral diversity of grasslands and arable field margins, for example, was mainly caused by the adoption of high-yielding forage crop varieties, increased fertilizer inputs, frequent applications of herbicides, and the increased purity of crop seed [7, 13]. Modern agricultural systems have produced a landscape in which many fields have very few weeds and very few invertebrates providing little food for birds. The shift in the type and density of weeds in the fields, as well as the disappearance of important habitats such as large stretches of hedgerows, was mainly responsible for the dramatic decline in bird populations [8, 14, 15]. Potential impacts of GM crops should thus be put in relation to the environmental impacts of modern agricultural practices that took place over the last decades.

1.2 Regulation of GM Crops

Generally, the approval of genetically modified crop varieties is more rigorously regulated than that of conventionally bred crops. Several reasons have lead to this regulation. The protection of human health and the environment is the primary reason for government oversight and regulation. There are other factors besides the safety aspect that have supported government decisions to regulate GM crops. Among others, there is the novelty of transgenic crops, the uncertainty accompanying the transformation process, and pub-

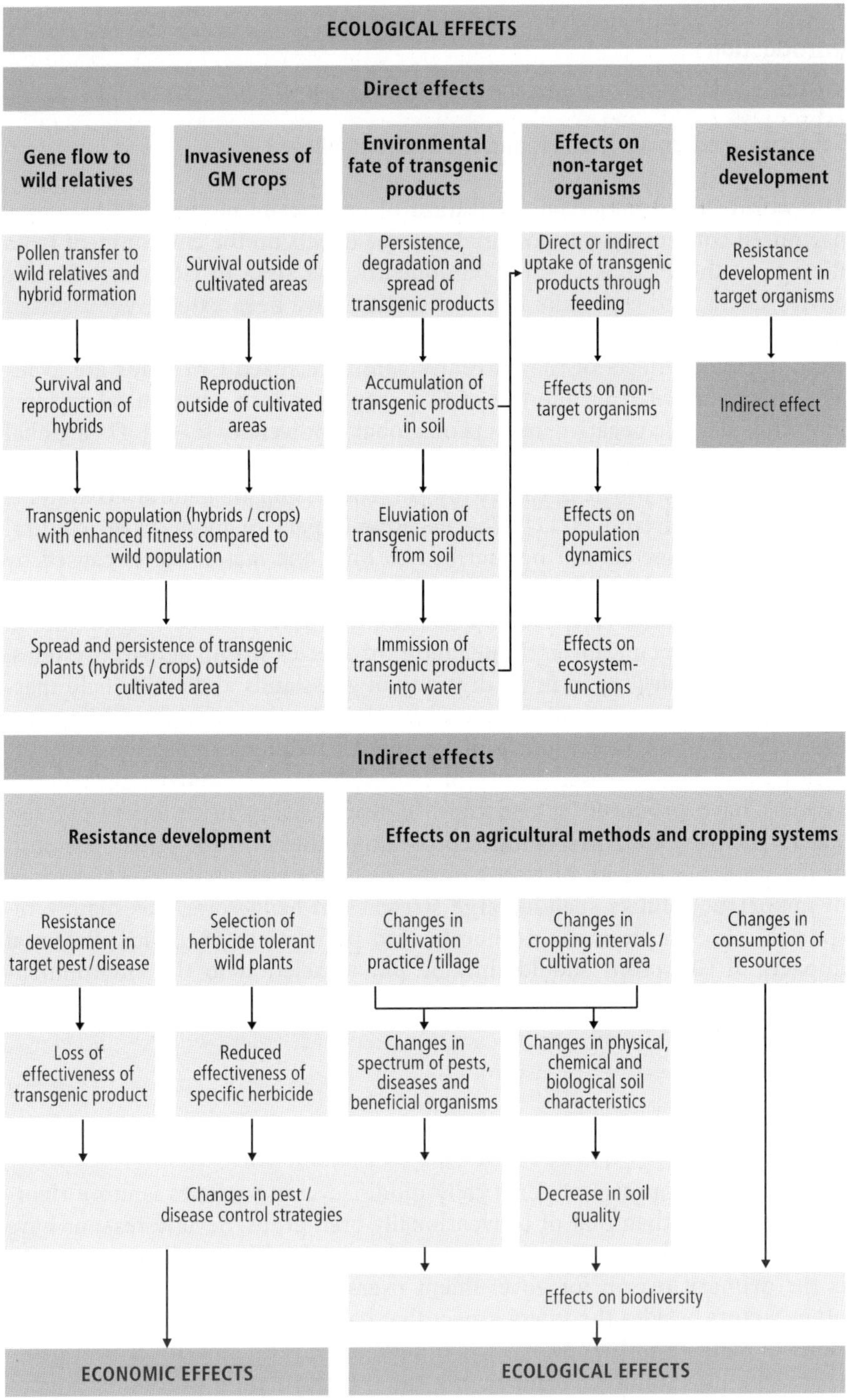
ECOLOGICAL EFFECTS
Direct effects
Gene flow to wild relatives
Invasiveness of GM crops
Environmental fate of transgenic products
Effects on non-target organisms
Resistance development
Pollen transfer to wild relatives and hybrid formation
Survival outside of cultivated areas
Persistence, degradation and spread of transgenic products
Direct or indirect uptake of transgenic products through feeding
Resistance development in target organisms
Survival and reproduction of hybrids
Reproduction outside of cultivated areas
Accumulation of transgenic products in soil
Effects on non-target organisms
Indirect effect
Transgenic population (hybrids / crops) with enhanced fitness compared to wild population
Eluviation of transgenic products from soil
Effects on population dynamics
Spread and persistence of transgenic plants (hybrids / crops) outside of cultivated area
Immission of transgenic products into water
Effects on ecosystem-functions
Indirect effects
Resistance development
Effects on agricultural methods and cropping systems
Resistance development in target pest / disease
Selection of herbicide tolerant wild plants
Changes in cultivation practice / tillage
Changes in cropping intervals / cultivation area
Changes in consumption of resources
Loss of effectiveness of transgenic product
Reduced effectiveness of specific herbicide
Changes in spectrum of pests, diseases and beneficial organisms
Changes in physical, chemical and biological soil characteristics
Changes in pest / disease control strategies
Decrease in soil quality
Effects on biodiversity
ECONOMIC EFFECTS
ECOLOGICAL EFFECTS

lic concerns about the safety of transgenic crops [16]. A thorough pre-market risk assessment of potentially unwanted effects of the GM crop on the environment is thus a prerequisite in obtaining permission to market any GM crop variety. GM crop growing countries generally follow the concepts of familiarity and of substantial equivalence, which state that a GM crop should be compared with its traditional counterpart that has an established history of safe use [17–20]. GM crop varieties that received regulatory approval are considered to present no more risks than comparable conventional varieties with a history of safe use.

1.3 Potential Environmental Effects of GM Crops

Potential environmental effects of the currently commercialized GM crops can roughly be subdivided into direct and indirect effects. Direct effects could result from the particular nature of the genetic change, i.e., from the resulting genotype and phenotype of the crop modified (Fig. 1). GM crops could be able to hybridize with sexually compatible wild relatives and these could subsequently suffer an increased risk of extinction. Introduced genetically modified traits could make a crop more likely to be more persistent (weedy) in agricultural habitats or more invasive in natural habitats. Transgenic products, especially toxins produced to be active against certain pests, could be harmful to organisms that are not intended to be harmed. Target pests could develop resistances against the insecticidal proteins produced in GM crops resulting in a loss of effectiveness of the transgenic product. Changes in the agricultural practice due to the adoption of GM crops (e.g., soil tillage, cropping intervals, or cultivation area) could result in a number of indirect effects (Fig. 1).

In the present review, the scientific knowledge of the environmental impact of GM crops deriving from 10 years of worldwide experimental field research and commercial cultivation is reviewed. The sources of information included peer-reviewed scientific journals, scientific books, reports from regions with extensive GM crop cultivation, as well as reports from international governmental organizations. The review is focussing on the currently commercially available GM crops that could be relevant for agriculture in Western and Central Europe (i.e., maize, oilseed rape, and soybean), and on the two main GM traits that are currently commercialized, herbicide tolerance (HT) and insect resistance (IR) [21]. Where helpful, experiences gained with other crops such as *Bt*-cotton are considered. GM crops with minor worldwide acreage (e.g., virus-resistant papaya and squash) are not considered. Potential effects

◀ **Fig. 1** Potential direct and indirect effects of genetically modified crops on the environment (adapted from [1, 2])

of GM crops are limited to the environment and to the following main topics: (1) effects of GM crops on non-target organisms, (2) effects of GM crops on soil ecosystems, (3) gene flow from GM crops to wild relatives, (4) invasiveness of GM crops into natural habitats, and (5) impacts of GM crops on pest and weed management. In addition, this review identifies the possible ecological benefits that could be derived from the cultivation of GM crops.

2 Effects of *Bt*-Crops on Non-target Organisms

Cry-proteins from *Bacillus thuringiensis* (*Bt*) are by far the most common insecticidal proteins that have been engineered into plants. They represent (up till now) the only insecticidal proteins that are commercially used in GM crops [21]. *Bt cry* genes have been engineered into a large number of plant species such as maize, cotton, potato, tomato, rice, eggplant, and oilseed rape [22–24]. However, at present, genetically modified *Bt*-maize and *Bt*-cotton are the only crops that are commercially cultivated. Transgenic *Bt*-potato plants expressing Cry3Aa to control the Colorado potato beetle (*Leptinotarsa decemlineata*) were commercialized from 1996 to 2001, but were withdrawn from the market due to lack of consumer acceptance and the introduction of a novel insecticide able to control both the Colorado potato beetle and aphids [24]. *Bt*-maize expressing Cry1Ab was initially developed to control a lepidopteran pest, the European Corn Borer (*Ostrinia nubilalis*), but has also shown to be effective against various other lepidopteran pests such as *Sesamia nonagrioides*, *Spodoptera littoralis* and *Helicoverpa zea* [25–27]. *Bt*-maize expressing the beetle-specific Cry3Bb toxin to control corn rootworms (*Diabrotica* spp.) has received commercial approval in 2003 in the United States and in Canada [28, 29]. However, due to its recent approval, no experience from commercial cultivation is yet available.

There are concerns that insect-resistant GM crops expressing Cry-proteins from *B. thuringiensis* could harm organisms other than the pest(s) targeted by the toxin. The long-term and wide-scale use of *Bt*-crops over the past 10 years has been accompanied by extensive studies testing potential adverse effects of these crops. One factor of particular interest in this respect is the potential effect of *Bt*-transgenic crops on non-target organisms that provide important ecological and economic services within agricultural systems. This includes parasitoids and predators that are of importance for natural pest regulation, pollinators, and butterflies.

2.1
Effects of *Bt*-crops on Beneficial insects (Predators and Parasitoids)

2.1.1
Lower-Tier Studies in the Laboratory and Greenhouse

The effects of *Bt*-crops on predators have been assessed in a number of studies, most of them using a tritrophic system including a plant, a herbivore and a natural enemy, i.e., predator or parasitoid (reviewed in [30]). Adverse effects on mortality, longevity or development of predators were only reported in studies using *Bt*-susceptible lepidopteran larvae as prey that had ingested the *Bt*-toxin. In particular, the green lacewing (*Chrysoperla carnea*), an important predator in many maize growing areas, has thoroughly been studied since studies suggested that this predator was negatively affected by Cry1Ab [31–33]. Results of subsequent studies using several different prey species reared on Cry1Ab-maize, however, showed that the insecticidal protein itself does not directly affect this predator, but that the green lacewing may be affected when feeding on prey species that are susceptible to *Bt*-toxin [34–36]. The negative effect observed was thus entirely prey-quality mediated, i.e., caused by the suboptimal food quality of the lepidopteran larvae used in the experiments. Because lepidopteran larvae are not considered an important prey for *C. carnea* in the field, it is unlikely that *Bt*-maize poses a risk for this predator [36, 37]. Similarly, effects of *Bt*-crops on mortality, development, weight or longevity of hymenopteran parasitoids developing in herbivores reared on transgenic plants were only observed in cases where *Bt*-susceptible herbivores were used as hosts [30]. This is not surprising given that host–parasitoid relationships are usually tight and parasitoids are very sensitive to changes in host quality. The results of the performed lower-tier studies provide evidence that except for the lepidopteran species the toxin is intended for, Cry1Ab does not cause direct toxic effects on any of the arthropod groups examined [30].

2.1.2
Higher-Tier Studies in the Field

More than 50 field experiments, varying greatly in size, duration, and sampling efforts, have been conducted to determine the effects of *Bt*-crops on natural enemies (reviewed in [30]). Most studies assessed the abundance of natural enemies using different methods, while only a few studies compared biological control functions of natural enemies in both *Bt*- and conventional crops. These experimental field studies have only revealed minor, transient or inconsistent effects of *Bt*-crops when compared to a non-*Bt* control [30, 38]. Indirect effects were observed with specialist natural enemies which were virtually absent in *Bt*-fields due to the lack of target pests as prey or hosts [39, 40]. Three studies in *Bt*-crops revealed consistent reductions in the abundance of different generalist predators that were also associated with the reduced availability of lepidopteran prey [41–43]. A 6-year field study in *Bt*-cotton on the abundance of 22 arthropod natural enemy taxa indicated that an average decrease of about 20% in some predatory species did not appear to be ecologically relevant for the biological control function of the natural enemy community [42, 44]. In general, many natural enemies are polyphagous, meaning they are able to switch to other preys in the field when one particular food source is scarce.

The occurrence of indirect effects that are caused by changes in the availability and/or the quality of target herbivores is not restricted to GM technology. Any pest-control measure will cause a reduction in the number of prey and host items, which could consequently affect population densities of natural enemies [30, 45, 46]. Such indirect effects are thus generally not considered to comprise a particular risk of insecticidal GM crops [20, 30].

A number of experimental field studies have included conventional insecticides as a treatment. Since *Bt*-crops are intended to replace or reduce applications of conventional insecticides commonly used in agriculture, insecticide treatments should be considered as one reasonable baseline for a comparative risk assessment [1, 3, 30]. Experiments that included broad spectrum insecticides, such as pyrethroids and organophosphates, have shown consistently reduced abundances of different groups of predators and hymenopteran parasitoids (*Bt*-maize [47–49]; *Bt*-cotton [42, 43, 50–53]). Side effects of more selective insecticides such as indoxacarb (anoxadiazine) or spinosad (amacrolide) largely depended on the spray frequency [49] whereas systemic insecticides (such as imidacloprid, a neonicotinoid) were found to have no or little effect on natural enemies [54]. Although some of the field studies were limited in their spatial scale, and lack statistical power due to limited replication and high variability in the data, they clearly indicated that non-target effects of *Bt*-crops were substantially lower than those of broad spectrum insecticides. This has been confirmed by recent large-scale studies conducted in commercially managed *Bt*- and non-*Bt*-cotton fields in the United States [55, 56]. The results of the various studies performed over the last years provide evidence that *Bt*-maize and *Bt*-cotton expressing insecticidal Cry1-proteins are more specific and have fewer side effects on non-target arthropods than most insecticides currently used.

2.2 Effects of *Bt*-crops on Pollinators

Many insect species are known to act as pollinators of various crops and wild plants. They are therefore of great ecological and economic importance. Among the various insect pollinators, honey bees are the best known, but it is now recognized that other species like bumble bees and solitary bees are also important in ensuring pollination of many plant species. Due to their ecological and economic importance, honey bees are often used as test species in pre-market risk-assessment studies to assess direct toxicity of insecticidal proteins on non-target organisms. Such studies have been conducted for each *Bt*-crop prior to its registration in the United States [57]. Feeding tests with Cry1Ab proteins were conducted on both honey bee larvae and adults and in each case no effects were observed [57]. Further studies with bees fed with purified *Bt*-proteins and with pollen from *Bt*-crops, as well as when bees were allowed to forage on *Bt*-crops in the field have confirmed the lack of effects [46, 58–60]

2.3 Effects of *Bt*-crops on Butterflies

Butterflies are considered as a species group with a high aesthetic value serving as symbols for conservation awareness. Since Cry1Ab is selectively toxic to Lepidoptera (moths and butterflies), off-site pollen flow from *Bt*-maize fields might potentially have adverse effects on Lepidopteran species, if their larvae feed on host plants dusted with *Bt*-pollen. The case of *Bt*-maize pollen and the monarch butterfly (*Danaus plexippus*) caused much public interest and led to a debate over the potential risks and the environmental impact of *Bt*-maize. Losey et al. [61] found that when pollen from a commercial variety of *Bt*-maize (event *Bt* 11) was spread on milkweed leaves in the laboratory and fed to monarch butterfly larvae, the larvae consumed significantly less from these leaves compared with leaves dusted with non-transgenic pollen. In addition, after 4 days, almost half of the tested larvae died, which was significantly more than on the leaves with non-transgenic pollen where none of the tested larvae died. The results of the study drew much attention to (potential) effects of *Bt*-crops on butterflies since the monarch is considered a conservation flagship species in the United States. However, the study also received much criticism and scientists questioned the validity of risk conclusions based on the data obtained in laboratory studies. Later laboratory bioassays showed that the only transgenic *Bt*-maize pollen that consistently affected monarch larvae was pollen from Event 176, an event that has meanwhile been withdrawn from the market. The results suggested that pollen from the most widely planted *Bt*-maize events (MON810 and *Bt* 11) will have no acute effects on larvae in field settings [62, 63] since their pollen expresses 80 times less toxin than Event 176 [63]. The results also suggested that pollen densities used by Losey et al. [61] were in excess compared to pollen densities present in maize fields or that the pollen of event *Bt* 11 used may have been contaminated with non-pollen tissues [64]. Excessive pollen densities of the currently commercialized events (*Bt* 11 and MON810) would be required to obtain relevant adverse effects on larval developments [62].

The critics also felt that in addition to the mere toxicity (hazard), an ecological risk assessment has to consider exposure, i.e., whether the monarch larvae will encounter the *Bt*-toxin and at what level. They also felt that the studies most likely did not address questions like the spatial and temporal overlap of monarch larvae and *Bt*-pollen. Extensive follow-up studies thus determined where the monarchs occur during their breeding season [65], and what percentage of the population of monarchs is possibly affected be the *Bt*-toxin in areas where *Bt*-maize is presently grown [66]. The results showed that larval exposure to pollen on a population-wide basis is low, given the proportion of larvae in maize fields during pollen shed, the proportion of *Bt*-maize fields, and the levels of pollen within and around maize fields [65]. The proportion of monarch butterfly population exposed to *Bt*-pollen was estimated to be less than 0.8% [66]. Field studies showed that continuous exposure of monarch butterfly larvae to natural deposits of *Bt*-pollen on milkweed plants within maize fields can affect individual larvae, but long-term exposure of larvae to *Bt*-maize pollen throughout their development is detrimental to only a fraction of the breeding population [67]. It was concluded that the risk of exposure is low and that it is unlikely that *Bt*-maize will affect the sustainability of monarch butterfly populations in North America [66, 67]. Furthermore, several authors claimed that effects of *Bt*-maize should be compared to mortality caused by other factors, which is very high in natural monarch butterfly populations, and averages around 80% over the entire larval development period [65, 67]. More important factors that may influence monarch butterfly survival include loss of over-wintering habitats in Mexico, use of insecticides to control lepidopteran pests and accidents such as collision with automobiles [57].

3 Effects of *Bt*-crops on Soil Ecosystems

Similar to non-target effects above ground, concerns were raised that *Bt*-crops could have effects on soil organisms and soil functions. The following section discusses the concern that non-target soil organisms and processes could be affected by the accumulation of *Bt*-toxins in soils through the cultivation of the currently commercialized *Bt*-crops.

3.1 Release, Persistence, and Biological Activity of *Bt*-toxins in Soil

Bt-toxins expressed in *Bt*-crops can enter the soil system either via root exudates, via senescent plant material, as well as via damaged and cast-off dead root cells [68–70]. The supply of *Bt*-toxins by senescent plant material mainly occurs via decaying biomass remaining on or in the ground after harvest. The toxin input from senescent plant tissue varies, depending on initial expression levels of the transgenic protein in different plant tissues, the progression of decay of the plant cells and the biomass remaining in the field. Expression levels in the *Bt*-maize variety MON810 are estimated to be around 4–7 times higher in leaves than in roots [71].

Persistence of *Bt*-toxins in soil is primarily depending on the protein quantity added and on the rate of inactivation and degradation by biotic and abiotic factors [72]. Degradation rates of *Bt*-toxins are known to be influenced by environmental conditions, soil type, the protein source (purified versus plant-produced) as well as by the particular Cry-protein chosen [45]. Persistence in the environment can be expressed in different ways, which affects comparison between studies. Terms such as dissipation time to 50% (DT50) or half-life are used to describe the time until 50% of the original amount of a substance is degraded. Persistence can also be described in terms of detectable residues. While, for example, a DT50 of 1–2 days is an indicator for a rapid rate of dissipation, detectable residues after 2–6 months indicate that some small amounts of the protein last in a biologically active form (if detected by a bioassay) or in an immunologically active form (if detected by ELISA). The description of detectable residues is a reference to an amount of substance that can be determined by an analytical method, but is not necessarily indicating biological activity. Determination of biological activity requires the use of an organism sensitive to the toxin [45].

Persistence, degradation, and inactivation of *Bt*-toxins have been assessed in the laboratory and/or in the field in 11 studies using either *Bt*-maize expressing Cry1Ab, *Bt*-cotton containing other Cry proteins or purified toxins (Table 1). The presented studies generally indicate an exponential degradation of *Bt*-toxins. After a short lag phase due to the breakdown of plant cells, a rapid degradation takes place with low amounts (< 2%) that may persist in soil after one season [70]. *Bt*-toxins may partially persist as a consequence of their binding to surface-active clay and humic acid compounds and it seems that bound proteins retain their insecticidal activity [69, 73–76]. To date, none of the laboratory or field studies suggest accumulation of *Bt*-toxins in soil over several years of cultivation. Experience from commercial cultivation indicates that *Bt*-toxin will not persist for long periods under natural conditions [72, 77, 78]. Although estimates on persistence of *Bt*-toxins differ among studies ranging from a few hours [79] to months [70, 80], the results are not essentially conflicting. Much of the described variation can be explained by the fact that the studies employ various parameters and experimental designs. In addition to environmental conditions varying between sites and seasons, degradation and persistence were depending on a multitude of factors including the type of *Bt*-toxin (Cry1Ab), the crop species (differences in C : N ratio), biotic activity (temperature), soil type (clay content), and the applied crop management practices (no-till with roots remaining in the soil).

Table 1 Summary of results from studies assessing persistence, degradation, and inactivation of *Bt*-toxins in soil

Bt-crop/ *Bt*-toxin	Study conditions	Toxin incorporation into soil	*Bt*-toxin detection	Persistence (days)	Refs.
Cotton tissue/ Cry1Ab and Cry1Ac	Laboratory	Experiments were carried out with field grown cotton tissue/soil/purified toxins in microcosms	Detectable residues (ELISA) [a]	Detection of toxin and insecticidal insecticidal activity at termination of test – 28 d (Cry1Ab) and 56 d (Cry1Ac)	[173]
Microbial toxin and cotton tissue/Cry1Ab and Cry1Ac	Laboratory	Purified toxin and transgenic leaves added to soil in microcosms. Toxins extracted and measured for 140 days	Detectable residues (ELISA)	Initial rapid degradation, low percentage may persist for weeks/ months. Half lives at 22/40 d, depending on clay/organic content of soil	[174]
Maize tissue/ Cry1Ab	Laboratory/ Greenhouse 24–27 °C	GM plants grown in greenhouse, harvest 2 weeks after pollen shed. Maize tissue was incubated with and without soil and mixed into artificial insect diet. Dose-weight response determined bioactivity. Soil: high clay content (25%)	Bioactivity test [b]	1.6 d (in soil) DT50 [c] 15.0 d (in soil) DT90 25.6 d (no soil) DT50 40.7 d (no soil) DT90	[175]
Cotton tissue/ Cry2A	Laboratory/ Field; Autumn/ winter MO, USA	Protein incubation in soil for 120 d. Bioassay based on growth inhibition to determine DT50	Bioactivity test	15.5 d (lab) DT50 31.7 d (field) DT50 120 d: down to < 25% (field&Laboratory)	[92]
Maize tissue/ Cry1Ab	Laboratory and field. Includes period of frost	Rhizosphere soil sampled from *Bt*-maize in a plant growth room and in the field	Western blot Bioactivity test	180 d: *Bt*-toxin detectable in rhizosphere soil samples from field (after first frost) around plants that had been dead for several months	[73]
Microbial toxin/Cry1F	Laboratory 25 °C	Mixture of Cry1F pipetted onto soil samples representative of cotton fields	Bioactivity test	< 1 d DT50	[79]
Bt-cotton cultivation/ Cry1Ac	Field ~ 16 °C	Soil samples were collected 3 months after post harvest tillage for 3–6 consecutive years	ELISA Bioactivity test	Not detectable *Bt*-toxins in any of the samples	[77]

Table 1 (continued)

Bt-crop/ *Bt*-toxin	Study conditions	Toxin incorporation into soil	*Bt*-toxin detection	Persistence (days)	Refs.
Maize tissue/ Cry1Ab	Litter bags in field (CH) ~ 9 °C	Leaves (growth chamber) sampled before/after pollen shed, cut&dried and placed in litter bags (5 mm mesh) and buried in soil in mid-October. Monthly analysis.	ELISA Bioactivity test	45 d DT50 145 d DT90 240 d : < 1.5% No degradation in winter (< 5 °C)	[70]
Maize tissue/ Cry1Ab	Soil cages in field (CH) ~ 9 °C	Leaves sampled 3 weeks after pollen shed, cut&dried and added to surface of soil cages (1 mm mesh) with earthworm, tied up in field for 200 d, starting December	ELISA Bioactivity test	35 d DT50 105 d DT90 200 d: 0.3% Degradation continued in winter	[70]
Maize tissue/ Cry1Ab	Laboratory and field. No temperature indication	Laboratory: *Bt*-maize residues added to soil and incubated for 43 days. Field: soil samples from experimental fields after 4 years cultivation of *Bt*-maize	ELISA	Laboratory: 14 d: Cry1Ab not detectable Field: most *Bt*-toxin in subsurface soil at 0–15 cm depth. Not clear if *Bt*-toxin from previous year	[78]
Maize tissue/ Cry1Ab	Field. No temperature indication MO, USA	After ≥ 3 years commercial cultivation of *Bt*-maize, soil samples were collected during growth period and 6 weeks after harvest. Growth inhibition determined presence of toxin	Bioactivity test	No evidence of persistence or accumulation	[72]
Maize tissue/ Cry1Ab	Field. No temperature indication Germany	Samples were taken during a 3-year monoculture study with MON810 from bulk and rhizosphere soil at a) 9 leaves per plant, b) stem elongation phase, c) flowering/anthesis, d) ripening	ELISA	No accumulation during growing season despite potential binding to soil particles. Proportion of toxin persisted through winter but no indication of accumulation, toxin in rhizosphere remained consistently higher than in bulk soil	[68]

[a] ELISA: Enzyme-Linked Immunosorbent Assay
[b] Bioactivity test: sensitive insect bioassay
[c] DT^{50}: Dissipation time 50% = time required for one-half of the initial quantity or concentration to dissipate from a system

3.2
Effects of *Bt*-crops on Soil Microorganisms

To date, the effects of *Bt*-crops on microorganisms have been evaluated in a number of studies which have used a range of different parameters and techniques [81, 82]. Most studies detected some differences when comparing *Bt*- with non-*Bt*-maize, however, the use of a wide variety of techniques makes a comparison among studies difficult [81]. The reasons for the observed differences as well as their implications are usually not clear. One difficulty in evaluating these changes is the high number of species in microbial soil communities and the natural variability occurring therein. In addition, the species and functional diversity of microbial soil communities is influenced by a multitude of environmental factors including plant species, water stress, fertilization, field management, tillage, fungal disease, grassland improvement, nitrification and soil depth [83]. Knowledge of the complex diversity of soil microorganisms is limited, since only a small portion of soil microbial populations can be cultured and identified using standard analytical methods [84]. Due to this limited knowledge, the importance and the functional consequences of detected differences in soil microbial populations are difficult to determine. Some methodological approaches, including the use of molecular biological techniques, show some promise in helping to understand the impact of GM crops on soil microbial ecology [81]. These molecular techniques yield fingerprint-type data, which represent an image of the soil microbial community analyzed [82, 85]. An accepted definition of the taxonomic unit, which can be used for defining soil microbial diversity, is, however, clearly lacking [85]. Because most studies assessing effects of GM crops on soil ecosystems have not determined the natural variation occurring in agricultural systems, it is generally difficult to establish whether the differences between *Bt*- and non-*Bt*-crops were exceeding this variation. The only study considering natural variation suggests that observed differences between *Bt*- and non-*Bt*-crops were not as large as differences caused by environmental parameters or by agricultural practices [86].

3.3
Effects of *Bt*-crops on Soil Macroorganisms

Effects of *Bt*-crops on soil macroorganisms have been investigated with nematodes, woodlice, springtails, soil mites and earthworms. Effects of Cry1Ab toxins on nematodes were examined in three studies using soil samples from fields planted with *Bt*-maize and non-*Bt* isolines [86–88]. The differences caused by the cultivation of *Bt*-maize were not as large as those resulting from cultivating different conventional maize cultivars, different crop plants, or as large as the differences between sites or sampling dates. The authors concluded that the effects found in *Bt*-maize fall within the normal variation expected in agricultural systems [86].

Three laboratory studies have shown that *Bt*-maize expressing Cry1Ab has no deleterious effects on the woodlice *Porcellio scaber* [89–91]. Wandeler et al. [91] compared six non-*Bt*-maize varieties and two transgenic *Bt*-maize varieties during a 20-day feeding experiment in the laboratory with regards to consumption by *P. scaber*. The consumption of maize leaves differed between the eight maize varieties. While *P. scaber* was found to feed significantly less on one of the two *Bt*-varieties compared to its corresponding non-transgenic control variety, the second transgenic variety was found to be one of the most consumed maize varieties when compared among all eight maize varieties evaluated. These results suggest that consumption by *P. scaber* was more strongly influenced by differences among the maize varieties used than by the factor *Bt*-variety alone.

No negative effects of the *Bt*-toxin Cry1Ab on two springtail species (*Folsomia candida* and *Xenylla grisea*) and on the mite species *Oppia nitens* were found in two laboratory studies [92, 93]. In addition, pre-market risk-assessment studies submitted for regulatory approval of several *Bt*-maize and *Bt*-cotton varieties have not revealed any toxic effect of Cry1A proteins on *F. candida* [57].

Effects of *Bt*-maize expressing Cry1Ab on the earthworm *Lumbricus terrestris* have been studied in the laboratory and under semi-field conditions in two studies [88, 94]. Both studies showed no consistent effects on *L. terrestris*. No significant difference in mortality and in weight of earthworms was found after 40 days in soil planted with *Bt*- or non-*Bt*-maize, or after 45 days in soil amended with the biomass of either *Bt*- or non-*Bt*-maize [88]. Laboratory experiments with adult earthworms feeding on *Bt*- and non-*Bt*-maize litter showed no significant difference in relative weight between the two treatments during the first 160 days of the experiment [94]. After 200 days, the authors found a significant weight loss of 18% of their initial weight when fed on *Bt*-maize litter compared to a weight gain of 4% of the initial weight of non-*Bt*-maize litter-fed earthworms. They concluded that further studies were necessary to see whether or not this difference in relative weight was due to the *Bt*-toxin. Under semi-field conditions, no significant differences in growth patterns were observed in immature *L. terrestris* feeding on *Bt*- and non-*Bt*-litter [94]. Pre-market risk-assessment studies submitted for regulatory approval have not revealed any toxic effect of Cry1A proteins on the earthworm *Eisenia fetida* [57]. In a recent study, the effects of *Bt*-maize on important life-history traits of the widespread earthworm *Aporrectodea caliginosa* were investigated under various experimental conditions [95]. Finely ground *Bt*-maize leaves added to soil had no deleterious effects on survival, growth, development or reproduction in *A. caliginosa*, even in high concentrations that could be considered as a worst-case scenario. Also, growth of juvenile *A. caliginosa* was unaffected when worms were kept in pots with a growing *Bt*-maize plant. The study confirmed the findings of earlier studies performed with other earthworm species [88, 94]. *Bt*-maize apparently poses minimal risks to earthworms as far as growth and reproduction is concerned.

3.4 The Ecological Significance of Effects of *Bt*-crops on Soil Ecosystems

Neither laboratory nor field studies have shown lethal or sublethal effects of *Bt*-toxins on non-target soil macroorganisms such as earthworms, springtails, soil mites, woodlice or nematodes. For soil microorganisms, many of the studies referred to in this section have focused on the detection of differences between *Bt*- and non-*Bt*-crops and they have been able to detect some differences in the number of species and in the composition of microbial soil communities. The limited knowledge on the complex diversity of soil microorganisms does, however, not allow to determine the importance and the functional consequences of detected differences in soil microbial populations. It is thus not possible to put an ecological value on these differences. To date, no evaluation has yet been published on the ecological relevance of differences in populations, communities or processes in soil ecosystems due to the cultivation of GM crops. With the exception of Griffiths et al. [86], observed differences have barely been compared with natural background variation, differences between conventional cultivars and crop systems, and impacts caused by routine pesticide application. In addition, knowledge gaps on the natural background variation occurring in agricultural systems still hinder the full interpretation of study results, making it difficult to clearly define what is considered an ecologically relevant effect on soil ecosystems. A final conclusion cannot be drawn, however, the scientific data obtained so far suggest that the effects owing to the cultivation of *Bt*-crops fall within the normal variation expected in agricultural systems. These variations are not as large as those resulting from growing different, conventional maize cultivars, crops, or as large as natural differences between sites or sampling occasions [86].

4 Gene Flow from GM Crops to Wild Relatives

The exchange of genes between crops and their wild relatives has always occurred, ever since the first plants have been domesticated. Natural hybridization of crops and related plants is considered to have played an important role in both domestication of crops and the evolution of weeds [3]. Surprisingly, gene flow from crops to wild relatives has only recently received major attention in the context of genetically engineered crops. Concerns have been raised that transgenes engineered into crops could be unintentionally introduced into the genomes of their free-living wild relatives [96]. Two major concerns related to transgenes in natural populations will be addressed in this section:

1. Could transgenes confer a benefit to weedy relatives (resulting in the evolution of so-called "superweeds"), which could then become very difficult to control in an agricultural environment? Weedy relatives are species related to crops which may grow within the crop or may occur in peri-agricultural environments, such as field margins or road verges.
2. Could wild relatives growing in "natural" environments suffer an increased risk of extinction due to hybridization with GM crops? Transgenic hybrids could become more competitive than the wild type (e.g., clover, alfalfa, and grasses). This would then lead to the extinction of the "wild-type" occurring outside arable agriculture in semi-natural habitat-types such as grass- or woodland.

It is generally agreed that the hazards related to gene flow from GM crops are linked to the introgression of transgenes into populations of wild relatives [1, 3, 97–99]. There is little scientific support for the assertion that transgene dispersal is a hazard in itself. This matter will therefore not be specifically addressed in this review.

4.1 Principles of Gene Flow

Transgene dispersal is often simply seen as pollen flow from the GM crop to its relative. The process of introgression, however, is not this simple, and actually occurs in many steps involving several hybrid generations [99]. Gene flow can roughly be separated into two processes: hybridization and introgression. For hybridization to occur, the transgenic crops and wild plants must grow within pollen dispersal distance, be sexually compatible, flower at the same time and viable pollen must be delivered to the stigma. Successful fertilization of the embryo must then be followed by zygote and seed formation. Introgression requires the hybrid seed to germinate and the first filial generation (F_1) plant to establish and flower in order to further hybridize with members of the recipient population [99, 100]. F_1 hybrids must therefore persist for at least one generation and be sufficiently fertile to produce backcross hybrids. Finally, backcross generations must progress to the point at which the transgene is incorporated into the genome of the wild relative.

Apart from the various biological factors mentioned, another important element determining the likelihood of transgene introgression is the occurrence of related species in the area where the crop is grown. Since most crops have been bred from wild plants it is not surprising that on a global scale nearly all crops may hybridize with a wild relative in some part of their distribution range [100]. However, only a small fraction of the world's flora has been domesticated and in modern agricultural systems, many crops are grown outside the range of the wild relatives with which they might hybridize [101]. The potential for gene flow from a specific crop therefore varies from region to region. In the following section, oilseed rape (OSR) (*Brassica napus*) is chosen as an example given that this is currently the only crop where GM varieties are widely commercialized and where gene flow to wild relatives must be considered in Switzerland [102].

4.2 Fitness of Transgenic Hybrids

The key issue whether a weedy plant might evolve to a more competitive weed after hybridization with a related GM crop or whether a transgene might increase the competitiveness of wild relatives in natural ecosystems depends on two factors: (1) does the transgenic trait confer a selective advantage to the wild plant, and (2) is the trait able to subsequently establish in a natural population. Fitness consequences of transgenes are therefore essentially depending on the character of the transgenic trait. The presence of a transgene does not in itself appear to be generally beneficial or detrimental in hybrids [96, 98]. The relative fitness of hybrids is depending both on the genotype and on the environmental conditions the hybrids are encountering. Transgenes that produce insect resistance (IR) will vary in their fitness potential—the common conclusion is that the transgenes will only confer a selective advantage if the fitness of wild populations is influenced by insect herbivores [98, 99]. Some studies were able to confirm this hypotheses, e.g., F_1 hybrids of oilseed rape and *Brassica rapa* containing *Bt*-genes were found to have a fecundity advantage under high insect herbivore pressure [103, 104]. However, these experiments also suggested that, in the absence of herbivores, fitness costs occur, which consequently are negatively influencing the competitiveness of the transgenic hybrids [98]. In most studies investigating the performance of transgenic hybrids between agricultural weeds and GM crops in semi-wild conditions, the hybrids were produced by artificial hybridization, i.e., they were crossed by hand pollination. Since many of these studies additionally manipulated environmental conditions, it is difficult to judge how hybrids would behave under natural conditions [98].

4.3 Hybrids of Oilseed Rape Becoming More Competitive Weeds in Agricultural Habitats

Commercial cultivation of oilseed rape (OSR) is to date the only situation that could possibly lead to the introgression of herbicide-tolerant genes into weedy relatives in Western and Central Europe. Examples of weedy relatives of OSR include wild turnip (*Brassica rapa*), wild mustard (*Sinapis arvensis*) and charlock (*Raphanus raphanistrum*). Any transfer of herbicide tolerance to these cruciferous weeds could render their control more difficult in both oilseed rape and subsequent crops in a rotation. Farmers would then have to find an alternative herbicide or a new control method.

Spontaneous hybrids between OSR and *B. rapa* are known to occur under field conditions with either species as the pollen donor [105–110]. However, the transfer of herbicide-tolerant genes from OSR to *B. rapa* seems to vary considerably in agricultural environments (Tables 2, 3). To date, only two studies have discovered herbicide resistant F_1 hybrids between *B. rapa* and OSR under commercial agricultural cultivation conditions [105, 110]. In a Canadian study conducted in Quebec, mean hybridization rates in feral populations of *B. rapa* were found to be 13.6% when sampled in or near a commercial field and 7% when sampled in two field experiments [110]. The higher frequency in commercial fields was explained to be most likely due to greater distances between individual *B. rapa* plants leading to higher pollen competition with OSR pollen. In contrast, in a similar study conducted during the Farm Scale Evaluations (FSE) in the UK, weedy *B. rapa* growing amongst OSR fields and within a 10-m strip next to the crop edge had been sampled, and only two out of approximately 9500 seedlings were found to have incorporated the herbicide-tolerant gene [105]. The considerable differences in the hybridization rates found in the two studies have not been elucidated yet. They could possibly be due to several factors:

- variations in the agricultural practice resulting in different amounts of *B. rapa* volunteers occurring as agricultural weeds
- variations in the fertility of the OSR cultivars used (conventional varieties vs. varietal associations) resulting in different amounts of transgenic pollen
- variations in the coincidence of flowering between both *B. napus* and *B. rapa*

The probability of gene flow from OSR to *S. arvensis* [111] and *R. raphanistrum* [112–114] seems to be very low (Tables 4, 5). The occurrence of spontaneous hybrids in commercial fields is therefore unlikely [105, 110].

4.4 Transgenic Hybrids Outcompeting Wild Types in Natural Habitats

To date, no long-term introgression of transgenes into wild populations leading to the extinction of any wild taxa has been observed [96, 98, 99]. Hybridization-mediated environmental impacts from the currently commercialized GM crops seem not to be any different from those of traditionally bred crops. However, transgene escape into wild populations of creeping bentgrass (*Agrostis stolonifera*) from experimental fields of GMHT creeping bentgrass has recently been demonstrated in the U.S. [115]. The long-term fate and ecological impacts of these transgenes within wild *A. stolonifera* pop-

Table 2 Summary of studies assessing gene flow from oilseed rape (*Brassica napus*) to wild turnip (*Brassica rapa*): assessment of fitness consequences using hybrids produced by artificial hybridization

Trait/Cultivar	Hybrid generation(s)	Experimental conditions	Method/marker used to confirm hybrid status[b]	Assessed fitness parameters	Hybridization (H) Fitness consequences (F)	Refs.
Herbicide-tolerant (HT) Oilseed rape (OSR) Glufosinate (Glu)	(F_1, BC_1)[a] (BC_2)	Experimental field trial	Herbicide spray, morphology, ploidy level	Pollen viability	H: 42% of the BC_2 plants obtained were Glu-tolerant F: Pollen fertility of BC_1 was greater than 90%	[176]
Non-transgenic OSR (cvs. Drakkar, Topas, Westar)	F_1	Experimental field trial	n.d.	Seed development, survival in the field, pod- and seet set	H: No strong hybridization barrier between *B. napus* and *B. rapa*. F: F_1-hybrids under some conditions nearly as fit as parents	[177]
Non-transgenic OSR (cvs. Topas, Westar)	F_2, BC_1	Experimental field trial	n.d.	Seed development, survival in the field, pod- and seet set	F: Relatively low average fitness of F_2 and BC_1 as compared to parents	[178]
HT OSR (Glu)	BC_3	Growth chamber	PCR, Herbicide spray	Pollen fertility, seed set, survival	F: No significant differences between transgenic and non-transgenic plants in survival and number of seeds per plant. Costs associated with transgene probably negligible	[179]

Table 2 (continued)

Trait/Cultivar	Hybrid generation(s)	Experimental conditions	Method/marker used to confirm hybrid status[b]	Assessed fitness parameters	Hybridization (H) Fitness consequences (F)	Refs.
Bt OSR	BC_1, BC_2	Growth chamber	PCR, Western Blot, ploidy level	n.d.	H: *Bt* trangene was present in hybrids and protein was synthesized at similar levels as corresponding OSR lines F: Not all F_1 lines were able to produce BC_1, but surviving BC_1 were able to produce BC_2	[106]
HT OSR (Glu)	F_1	Experimental field trial	Morphology, AFLP, PCR	Flower, pollen and seed production	F: Male fitness among F_1 produced by *B. rapa* is low	[180]
Bt/GFP OSR	F_1, BC_1, BC_2	Experimental field trial	GFP	Vegetative plant material produced in an insect bioassay	F: No difference found in biomass between BCs and non-transgenic parents under low insect pressure	[103]
OSR	F_1, F_2, sev. BCs	Experimental field trial	n.d.	Seed production	F: Hybrids are not generally less fit than parents. Fitness of both parents and hybrids is strongly frequency-dependent	[181]

Table 2 (continued)

Trait/Cultivar	Hybrid generation(s)	Experimental conditions	Method/marker used to confirm hybrid status[b]	Assessed fitness parameters	Hybridization (H) Fitness consequences (F)	Refs.
Bt/GFP OSR	F_1	Green house	GFP	Biomass, flower number, seed mass, germination rate	F: Herbivore pressure and plant density had strong impact on relative biomass and on fitness advantages of *Bt*-hybrids over wild type. Greenhouse results cannot give a quantitative prediction of *Bt*-spread and persistence in natural habitats	[104]
Bt/GFP OSR	F_1, BC_1, BC_2	Experimental field trial	GFP	Intraspecific competition with various herbivore pressures and with wheat	F: On average hybrids of various BC generations have lower potential for growth and competitiveness under field conditions than weedy parents	[182]
Male-sterile OSR	F_1, BC_1	Growth chamber Experimental field trial	Quantitative PCR	Photosynthetic capability, pollen viability, seed set	H: Expression of transgenes is stable in F_1 hybrids. F: Reproductive fitness of hybrids was significantly lower than in parents, BC_1 had significant lower photosynthetic capability and reproductive fitness than parents. Vegetative vigor of of BC_1 is limited.	[183]

[a] Hybrids of F_1 and BC_1 generations used in this study were produced under natural hybridization conditions; F_1= first filial generation, F_2= second filial generation, BC_1= first backcross generation etc.
[b] *GFP* green fluorescent protein; *PCR* Polymerase chain reaction; *AFLP* Amplified fragment length polymorphism; *RFLP* Restriction fragment length polymorphism; *n.d.* not determined

Table 3 Summary of studies assessing gene flow from oilseed rape (*Brassica napus*) to wild turnip (*Brassica rapa*): assessment of hybridization rates under natural hybridization conditions

Trait/Cultivar	Hybrid generation(s) [a]	Experimental conditions	Method/marker used to confirm hybrid status [b]		Hybridization (H) Fitness consequences (F)	Refs.
Non-transgenic oilseed rape (OSR) (cv. Drakkar)	F_1	Agricultural field (set-aside)	AFLP	H:	First study to show introgression between *B. napus* and *B. rapa* under natural condition. Hybrids in weedy natural populations resembled most closely to BC_2 (obtained by controlled crosses)	[108]
Bt OSR	F_1	Experimental field trial	Antibiotic marker	H:	F_1 hybrids have similar levels of expression as crop lines (when hybridization occurs under natural conditions)	[106]
Herbicide-tolerant (HT) OSR Glyphosate (Gly)	F_1	Experimental field trial	Herbicide spray, Gly test strip, ploidy level, AFLP	H:	Hybridization between *B. napus* and *B. rapa* occurred at approx. 7%	[110]
HT OSR (Gly)	F_1	Commercial field	Herbicide spray, Gly test strip, ploidy level	H:	Hybridization between *B. napus* and *B. rapa* occurred at approx. 13.6%	[110]
GFP OSR	F_1	Experimental field trial	GFP, morphology, pollen viability ploidy level	H:	Hybridization between *B. napus* and *B. rapa* occurred at approx. 7%	[110]
OSR	F_1, BC_1	Agricultural field (set-aside)	Chromosome counting, AFLP	H:	Introgression progresses primarily with *B. rapa* as maternal plant. Transgenes can be transferred from *B. napus* to *B. rapa*	[109]
Bt/GFP OSR	F_1, BC_1	Experimental field trial	GFP	H:	Hybrids between *B. napus* and *B. rapa* occurred over a wide range of experimental conditions, BC_1 rate was 0.074%	[107]
HT OSR (Glu)	F_1	Agricultural field	Herbicide spray, PCR, ploidy level	H:	2 hybrids found in 9500 seedlings	[105]

[a,b] Abbreviations: see Table 2

Table 4 Summary of studies assessing gene flow from oilseed rape (*Brassica napus*) to charlock (*Raphanus raphanistrum*)

Trait/Cultivar	Hybrid creation/ generation(s) [a]	Experimental conditions	Method/marker used to confirm hybrid status [b]	Fitness parameters used	Hybridization (H) Fitness consequences (F)	Refs.
Male-sterile oilseed rape (OSR) cv. Brutor	N F_1, F_2, BC_1	Experimental field trial		Seed production	H: Hybrid frequency expected to be at max. 0.2%. Seed production of F_1 = 0.4%, F_2 = 2%	[112]
Non-transgenic OSR (Acetolactat synthase-resistant)	N F_1	Experimental field trial	Morphology, RFLP, ploidy level	Pollen viability	H: No hybrids were detected amongst 25 000 seedlings collected from *R. raphanistrum*. Two hybrids were detected in more than 52 Mio. OSR seedlings. F: Both hybrids had viable pollen and were able to set seed when backcrossed to *R. raphanistrum*, but not OSR	[184]
Herbicide-tolerant (HT) OSR (Glu)	N BC_6	Experimental field trial	Herbicide spray, PCR, ploidy level	Seed production and survival, plant growth and reproduction	H: n.d. F: Fitness level of backcrosses with OSR is 100× lower than of BC with *R. raphanistrum*.	[114]
OSR	N F_1	Experimental field trial	Morphology, ploidy level	Seed emergence, flowering time and frequency, diameter of rosette, dry weight	H: n.d. F: F_1 hybrids showed lower seedling emergence, significant delay of emergence and lower survival than both parents	[113]

Table 4 (continued)

Trait/Cultivar	Hybrid creation/ generation(s) [a]	Experimental conditions	Method/marker used to confirm hybrid status [b]	Fitness parameters used	Hybridization (H) Fitness consequences (F)	Refs.
HT OSR (Gly)	A F_1	Green house	Herbicide spray, AFLP, ploidy level	n.d.	H: No hybridization detected F: n.d.	[110]
HT OSR (Gly)	N F_1	Experimental field trial	Herbicide spray	n.d.	H: One hybrid detected in approx. 32 000 seedlings F: n.d.	[110]
HT OSR (Gly)	N F_1	Commercial field	Herbicide spray	n.d.	H: No hybridization detected F: n.d.	[110]
OSR (GFP)	N F_1	Experimental field trial	GFP	n.d.	H: No hybridization detected F: n.d.	[110]
Bt-OSR containing GFP	N F_1, BC_1	Experimental field trial	GFP	n.d.	H: No hybridization detected F: n.d.	[107]

[a] A = Hybrids were produced by artificial hybridization (e.g. hand-pollination), N = Hybrids produced under natural hybridization conditions

[b] Abbreviations: see Table 2

Table 5 Summary of studies assessing hybridization rates between oilseed rape (*Brassica napus*) and wild mustard (*Sinapis arvensis*) and dog mustard (*Erucastrum gallicum*) [a]

Trait/Cultivar	Hybrid creation/ generation [b]	Experimental conditions	Method/marker used to confirm hybrid status [c]		Result	Refs.
Six non-transgenic oilseed rape (OSR) cultivars	A/N F_1	Green house Experimental field trial	PCR, Morphology, Southern blot	H:	Neither *S. arvensis nor B. napus* readiliy hybridise with each other in the Greenhouse. Unable to detect gene flow from *B. napus* to *S. arvensis* in the field	[111]
Herbicide-tolerant (HT) OSR (Gly)	N F_1	Commercial field	Herbicide spray	H:	No hybridization detected	[110]
HT OSR (Glu)	N F_1	Agricultural field	Herbicide spray, PCR	H:	1 hybrid found in the field	[105]

[a] *E. gallicum* was only investigated in Warwick et al. 2003
[b] A = Hybrids were produced by artificial hybridization (e.g., hand-pollination), N = hybrids produced under natural hybridization conditions
[c] Abbreviations: see Table 2

ulations remain to be determined. Gene flow from traditional crops has on some occasions created problems by bringing wild relatives closer to extinction. There are two known examples of crop-gene flow that have led to the evolution of decreased fitness in wild populations. Natural hybridization of an endemic wild rice species (*Oryza rufipogon* ssp. *formosana*) with cultivated rice (*Oryza sativa*) contributed to its extinction in Taiwan [96]. Similarly, genetic pressure due to the cultivation of the purple flowering alfalfa (*Medicago sativa*) has lead to the disappearance of the yellow flowering wild-type (*M. falcata*) from large areas in Switzerland [116].

4.5
Conclusions on Gene Flow to Wild Relatives

There is general agreement that gene flow from GM crops to sexually compatible wild relatives can occur. Experimental studies have shown that GM crops are capable of spontaneously mating with wild relatives, however, at rates in the order of what would be expected for non-transgenic crops [96]. Much empirical information about crop-wild relative hybridization is now available [97] indicating that such hybridization occurs when sexually compatible wild relatives are present in close proximity to the crop, albeit at low (and variable) rates [99]. Hybridization between conventional (non-GM) crops and their wild relatives has occasionally caused problems in ecological and evolutionary time. There is no evidence as yet that GM crops pose any greater risk than do non-GM crops, but our knowledge of the fitness consequences of transgenes in wild populations is incomplete [98]. It is difficult to judge a priori whether a transgenic phenotype will have a special fitness advantage relative to a non-transgenic counterpart—and if an advantage exists, whether this will result in increased weediness.

5
Invasiveness of GM Crops into Natural Habitats

The awareness of the problems that sometimes accompanied the deliberate or accidental introduction of non-native species into new environments has a long history [117]. Invasions have been recognized in a growing number of environments as being serious threats to the preservation of what we choose (by our choice of time scale) to be regarded as native fauna and flora [118–120]. Although the great majority of accidental introductions undoubtedly failed to become established, a substantial number became established, and some of these became serious pests [121]. Not surprisingly, the concern of GM crops invading natural habitats was brought up early in the discussion on potential environmental risk related to the release of GM crops [121].

5.1 Multiple Herbicide Resistances in Oilseed Rape Volunteers

Gene flow between different transgenic OSR growing in habitats which are frequently disturbed (such as road verges) has commonly been part of the discussion on environmental effects of GM crops, especially in Canada. There are three types of herbicide-tolerant OSR commonly grown in Canada: glyphosate (counting for 59% of the total acreage in 2001) and glufosinate-resistant varieties (16%)—both obtained by genetic engineering—as well as a non-transgenic imidazolinone-resistant type (25%) [122]. It was conceived that the transfer of herbicide-tolerance genes between varieties of OSR through gene flow may result in volunteers resistant to two or more herbicides, which could pose agronomic problems in volunteer plant control. After 3 years of commercial cultivation of GMHT OSR, two triple-herbicide resistant volunteers were reported at a field site in western Canada [123] and a study at 11 sites in Saskatchewan, Canada, reported double-resistant OSR volunteers [124]. The results of both studies suggest that HT gene stacking can occur in OSR volunteers. This is not surprising given the outcrossing potential of OSR, the large acreage of GMHT OSR in Western Canada, and the potential seed bank life leading to the incidence of OSR volunteers [122, 123, 125]. Rotations including many GMHT crops having the same trait (e.g., glyphosate tolerance) may result in various crop volunteers resistant to the same herbicide and thus make certain cropping systems fragile [125]. However, there is no evidence at present that the extensive cultivation of GMHT OSR over several years in western Canada has resulted in an increase of volunteer OSR that would have been caused by the herbicide-tolerant traits [126]. Extensive weed population monitoring has been conducted in thousands of fields and will continue to play an important role in assessing populations of herbicide-tolerant volunteers, weed population shifts, and changes to weed biodiversity due to GMHT crops. The lack of reported multiple-resistant volunteers suggests that these volunteers are being controlled by chemical and non-chemical management strategies, and are therefore not an agronomic concern to most producers [123, 126]. The multiplicity of herbicides available ensures that HT gene-stacked volunteers are not an agricultural problem. In Canada, there are over 30 registered herbicides to control single- or multiple-resistant GMHT OSR in cereals, the most frequent crop to follow OSR in a typical 4-year rotation [122]. In all crops, except field peas, alternative herbicides are able to control herbicide-tolerant OSR because glyphosate and glufosinate are not used in crops other than OSR at this time in western Canada [126]. Although not all volunteer OSR are killed by the herbicide application, most survivors are affected by the combination of crop competition and partial herbicide control that reduces seed set. Furthermore, there are a multitude of cultural and mechanical practices that are recommended to growers to manage multiple-GMHT OSR volunteers. These include [122] (1) leaving seeds on or near the soil surface as long as possible after harvest because a high percentage will germinate in the fall and be killed by the frost; (2) using tillage immediately before sowing; (3) silaging and green manuring to prevent seed set in volunteers; (4) isolating OSR fields with different HT traits; (5) following OSR with a cereal crop and rotating OSR in a 4-year crop rotation; (6) scouting fields for volunteers not controlled by weed management; (7) using certified seed and (8) reducing seed loss during harvest.

5.2
Invasiveness of Transgenic Crop Varieties into Semi-natural Habitats

Not many experimental studies have been performed comparing the invasiveness of transgenic crop varieties to non-transgenic varieties. In an early study, population dynamics of GMHT OSR with a resistance to glufosinate and conventional OSR were estimated over a 3-year period in 12 natural habitats and under a range of climatic conditions [127]. There was no evidence that genetic engineering for herbicide tolerance increased the invasive potential of OSR in undisturbed natural habitats. Furthermore, there was no evidence that transgenic OSR was more invasive or more persistent in disturbed habitats compared to their conventional counterparts. In general, the transgenic lines performed even less well than the non-transgenic lines. A more recent study compared four different crops (both conventional and GM) grown in 12 different habitats and monitored their performance over a period of 10 years [128]. In no case the GM crops (OSR and maize expressing tolerance to glufosinate, sugar beet tolerant to glyphosate, and two types of GM potato expressing either the *Bt*-toxin or a pea lectin) were found to be more invasive or more persistent than their conventional counterparts.

5.3
Conclusions on the Invasiveness of GM Crops Into Natural Habitats

Despite the extensive commercial cultivation of GMHT OSR in western Canada for several years, there is currently no evidence of GMHT OSR becoming feral. This is due to its lack of persistence in the seed bank, the redundant and repetitive control of volunteer weeds in subsequent crops, the absence of persistent populations in ruderal areas, and the limited occurrence of weedy relatives with a potential for hybridization [126]. De-domestication of crops and associated ferality appears to be restricted to only a few crop groups. They are only of minor importance globally with regard to invasive weed problems especially compared to other plant groups [129]. Globally, the feral plants that cause much of the economic damage are imported horticultural plants [118–120]. Unlike annual crops, these horticultural plants are mostly perennials that have extensive sexual and asexual reproduction.

6 Weed Management Changes Related to GM Herbicide-tolerant Crops

Environmental impacts due to crop management changes are usually difficult to assess because they are often caused by many interacting factors and do only show up after an extended period of time. Not surprisingly, the impacts of modern (non-GM) agriculture on biodiversity were only revealed years after these techniques had been introduced (see Sect. 1.1). Considering the widespread effects modern agricultural systems had in the last decades, changes in management practices are probably among the most influential factors that could lead to biodiversity changes. It appears that concerns related to crop management changes have been perceived more strongly and have been judged to be more important since the adoption of GM crops and that these concerns were less prevalent in the past.

6.1 Shifts of Weed Populations and Potential Impacts on Biodiversity

The impacts on farmland biodiversity due to the use of genetically modified herbicide-tolerant (GMHT) crops are currently discussed in two contrasting matters. While there are concerns that the control of weeds in GMHT crops using broad-spectrum herbicides might be so efficient that long-term declines in weeds could lead to the decline of wildlife depending on them [130, 131], others suggest that GMHT crops might ameliorate farmland biodiversity by delaying and reducing herbicide use, and even allowing weeds and associated wildlife to remain in fields longer [132–134].

The concern that declines in weed number could have adverse effects on farmland biodiversity received major public attention due to the interpretations of the results of the Farm Scale Evaluations (FSE) performed in the United Kingdom. The FSE were able to show that the biomass of weeds was reduced under GMHT management in sugar beet and oilseed rape and increased in maize compared with conventional treatments [135]. However, the invertebrate groups assessed (herbivores, detritivores, pollinators, predators and parasitoids) were much more influenced by season and by crop type than by the GMHT management [136]. The abundance of many invertebrate groups increased two-fold to five-fold between early and late summer, and differed up to 10-fold between crops, whereas GMHT management superimposed relatively small (less than twofold), but consistent, shifts in weed and insect abundance.

The results of the FSE led some to the rather simplistic conclusion that the use of GMHT crops generally leads to lower weed and insect densities, which consequently affect farmland biodiversity, and especially bird populations. Although the FSE were one of the most extensive ecological studies ever conducted, they were not without limitations [137, 138]. As the authors of the

FSE studies stated, "the FSE addressed one particular environmental risk of one particular trait in one particular agro-ecosystem, and the results should not be extrapolated to other socio-environmental systems" [139]. There are two important limits that we feel should be critically discussed:

Extrapolation of the Results from the Farm to the Landscape Level
The effects observed in the FSE were restricted to the field-scale. Taking into account that all three crops occupied less than 15% of the total arable field surface of Great Britain in any year [135], it is unclear if these effects would occur at the landscape-level and how significant they would be. A major factor in the decline in farmland biodiversity over the last decades has been the loss of more specialized taxa [8]. Thus, many of the birds and butterflies that declined markedly in the period prior to 1970 were dependant on areas of extensive low-input cultivation or the presence of non-cropped habitat. In general, the plants currently common on arable land are found in a wide range of other habitats. Similarly, butterflies as well as the non-declining farmland birds now typical of farmland in Britain are those that tend to be habitat generalists [8]. More intensive field management, degradation in habitat quality, and increasing habitat homogeneity (across all-scales) are currently the most important drivers of biodiversity loss.

Consequences of the Cropping and Weed Management System Applied
The FSE assumed that no other changes in field management will occur other than the GMHT crops replacing present non-GM varieties in a proportion of fields [135]. The results are therefore linked to the weed-management system practiced in the FSE, for both conventional and GMHT systems. Highly effective weed control practices such as those chosen for the GMHT crops in the FSE lead to low numbers of weed seeds and insects. In turn, fewer insects and decreased weed seed might reduce the numbers of birds that depend on these insects and seeds as a food source [137]. However, other weed-management systems than the one used in the FSE are possible. The use of GMHT technology in the U.S. and in Canada was accompanied by a series of management changes including the adoption of conservation tillage practices, which are considered to have several environmental benefits [140, 141] (see Sect. 7). These include beneficial impacts on farmland biodiversity, because conservation tillage results in a greater availability of crop residues and weed seeds improving food supplies for insects, birds, and small mammals [142]. Similarly, studies conducted in the UK have shown that alternative scenarios to those resulting from the FSE are possible for GMHT sugar beet [132, 134]. GMHT sugarbeet allows to choose an optimal application time and to reduce the number of herbicide sprays, resulting in environmental benefits compared with the conventional practice. Depending on the herbicide management chosen, it can either enhance weed seed banks and autumn bird

food availability, or provide early season benefits to invertebrates and nesting birds [134].

6.2
Selection of Resistant Weeds by Intensive Herbicide Applications

The wide adoption of GMHT crops raised concerns that the increasing applications of one herbicide will rapidly enhance the evolution of herbicide-tolerant weed populations. However, independently from the adoption of GM crops, a number of changes have occurred in conventional agricultural systems during the past decades, which resulted in significant impacts on weed communities. The most important selective forces on a weed community in a crop rotation system are tillage and herbicide regime. Most of the resistant biotypes evolved without the selection pressure resulting from the adoption of GM herbicide-tolerant crops. Numerous weed species have evolved resistance to a number of herbicides in many, if not most, agricultural systems long before the introduction of GMHT crops [143, 144]. The commercialization of herbicides inhibiting acetolactat synthase (ALS), for example, induced the evolution of herbicide-resistant biotypes in over 90 weed species, while 65 weed species have evolved resistance to atrazine [143, 144]. It seems that tolerance to glyphosate, in contrast, is less likely to develop in weed species (and in volunteers) than tolerance to other herbicides, as a result of its chemical properties and its mode of action [145, 146]. After almost three decades of glyphosate use, tolerance to glyphosate has only been reported in eight weed species worldwide [143].

The experiences available from regions growing GMHT crops on a large-scale confirm that the development of herbicide-resistance in weeds is not a question of genetic modification, but of the herbicide management applied by farmers. In Canada, no weed species have been observed yet that demonstrated herbicide tolerance to glyphosate [146]. Although no long-term studies have been conducted, no significant shifts in weed populations and no major difficulties in the management of weeds in agricultural settings have been attributed to the widespread cultivation of GMHT crops in Canada either. This is, in part, certainly due to farmers rotating both their crops and the herbicides they use for weed and volunteer control. In the United States, in contrast, glyphosate has been used before the introduction of GMHT varieties in combination, or in sequence with other herbicides in continuously cultivated no-tillage soybean fields. With the widespread use of GMHT soybeans, many fields have been treated only with glyphosate, which increased the pressure for the selection of resistant weed biotypes. As a consequence, within 3 years after the introduction of GMHT soybean varieties, glyphosate-resistant horseweed (*Conyza canadensis*) was detected [147]. It is clear that the continuous application of the same herbicide in one particular crop over multiple years without applying appropriate crop rotation will inevitably lead to the selection of herbicide-tolerant weeds. The limited number of herbicides used results in greater selection pressure on the weed community.

Glyphosate-resistant weeds have been described by some as "super weeds", and there have even been inferences that glyphosate-resistant weed presence could reduce farmland value. Although farmers have to add another herbicide to glyphosate to control the resistant weed species, there are alternatives to glyphosate that are highly effective and provide good flexibility in application timing for most weed species. There is, however, no question that glyphosate-resistant weeds will increase the costs of weed management to farmers. A more costly scenario would involve a weed for which the alternative herbicides have limited flexibility in application timing. In this situation, the loss of application flexibility would present a greater cost to many farmers than the additional herbicide expense.

In conclusion, the simplest way for farmers to reduce selection pressure placed on weeds by glyphosate is to avoid planting continuous glyphosate-resistant crops and to annually rotate the herbicides used. Such procedures are in fact part of any reasonable herbicide resistance management strategy that should be followed by farmers and that are recommended by regulatory agencies in Europe and in North America, as well as by the industry [148–150].

6.3 Changes in Herbicide use due to GMHT Crops

There are many criticisms arguing that the adoption of GMHT crops would generally lead to an increased use of herbicides. Studies can be found to support this view [151, 152], but there appear to be more studies that support a small but statistically significant reduction in herbicide use [140, 153–155]. Because the reduction varies between crops and regions, it is difficult to draw a general conclusion. The adoption of GMHT varieties of oilseed rape in Canada, for example, has been associated with a reduction in the amount of herbicide used per hectare as well as a decline in the potential environmental impact of chemical weed management [153]. The average soybean herbicide application rates in the U.S., in contrast, have slightly increased by 3% since the introduction of GMHT soybean (in terms of active ingredients per acreage) [140, 155]. It would, however, be insufficient to assess herbicide use only by comparing the quantities of herbicides applied, even if expressed as the total amount of active ingredient. Beside net changes in the amounts used, the adoption of GMHT crops has more precisely resulted in a change in the mix of herbicides used. The assessment of this change, however, is not as straightforward as it may seem, since toxicity and persistence in the environment vary across pesticides. Assessing herbicide changes relying purely on the amounts used, would assume that the same amount of any two ingredients has equal impact on human health and the environment, while in fact the various active ingredients in use in herbicides vary widely in toxicity and in persistence in the environment. The adoption of GMHT crops has allowed farmers to use herbicides (glyphosate and glufosinate) that are less toxic to humans and to the environment than the previously used [155, 156]. In some countries, especially in South America, the adoption of GMHT soybeans increased the volume of herbicides used relative to the amounts used before GMHT adoption [154, 157, 158]. This is largely due to the fact that the GMHT technology has accelerated the switch from a conventional tillage system (where no or less herbicides were used because weeds were mainly ploughed into the soil) to a conservation tillage system. The increase in the net volume of herbicides used should, however, be placed in the context of the environmental benefits of the new conservation tillage systems (see Sect. 7).

7 Possible Ecological Benefits of GM Crop Cultivation

7.1 Pesticide Reductions due to Insect-resistant Crops

Studies on the economic impacts of insect-resistant GM crops are revealing benefits for farmers, most of all where yields are hampered by high pest incidence or where the development of resistant pests impedes the use of pesticides [159, 160]. The benefits related to the adoption of *Bt*-crops may comprise both higher yields and significant reductions in pesticide use for some crops. While the adoption of *Bt*-maize expressing the insecticidal protein Cry1Ab has resulted in only modest reductions in insecticide applications due to the small area of conventional maize treated with insecticides, the commercial cultivation of *Bt*-cotton has proven to have resulted both in a significant reduction in the quantity and in the number of insecticide applications [159, 161]. Cotton is highly susceptible to several serious insect pests belonging to the budworm-bollworm complex, i.e., tobacco budworm (*Heliothis virescens*), cotton bollworm (*Helicoverpa* spp.) and pink bollworm (*Pectinophora gossypiella*). These insects constitute a major problem in most cotton-growing areas because they can cause considerable damage. Conventional cotton cultivation therefore relies heavily on repeated insecticide applications throughout the growing season. Although estimates on pesticide use vary because pesticide use is depending on regional pest pressures, management practices and yearly variations, it appears that the adoption of *Bt*-cotton has significantly reduced the numbers of pesticide applications in every country where *Bt*-cotton has been grown [161]. Moreover, most studies estimate a reduction in the amount of pesticides used [141, 154, 161]. Direct environmental benefits of reduced insecticide applications in *Bt*-cotton resulted in fewer non-target effects [55, 56] and in reduced pesticide inputs in water [159]. In China, for example, the number of pesticide applications against lepidopteran pests in cotton has considerably dropped from nine in 1994 to four applications in 2001 following the adoption of *Bt*-cotton [162]. Concerns have been raised that these environmental benefits may be lowered by additional spraying against secondary pests that were formerly controlled by the broad spectrum pesticides. There is, however, no published evidence that *Bt*-cotton has resulted in a general change in the pest spectrum leading to an overall increase of pesticide applications. In addition to direct environmental benefits, pesticide reductions related to the adoption of *Bt*-cotton have also shown to have reduced many immediate as well as longer-term risks to human health [163–166].

7.2
New Weed Control Strategies Offered by GM Herbicide-Tolerant Crops

The adoption of GMHT crop varieties has resulted in several weed management changes compared to conventionally managed crops. GMHT crop varieties allow the use of a single broad-spectrum herbicide that has a wider spectrum of activity and that may reduce the need for herbicide combinations or chemicals that require multiple applications [153, 155, 156]. The herbicides used in GMHT crops (glyphosate or glufosinate) are foliar-applied, post-emergence herbicides, which usually allow using herbicides in a more targeted manner. They can be applied after weeds have emerged, i.e., areas with high weed densities can be identified and treated, while areas with low weed pressure can be treated with reduced herbicide amounts. Post-emergence herbicides are thus generally applied at lower rates than soil-applied, pre-emergence herbicides, also because absorption by soil colloids and degradation are reduced [167]. Glyphosate and glufosinate are considered being less toxic to human health and the environment than many of the herbicides they replace [155, 156]. Both have relatively short soil half-lives and they persist almost half as long in the environment compared to the replaced herbicides. Neither moves readily to ground water, which results in fewer losses of chemicals by leaching and run-off from the field [156].

Perhaps the most important environmental benefit of the adoption of GMHT crops is the possibility to use broad spectrum herbicides, which encouraged growers to adopt conservation tillage strategies [140, 156, 168, 169]. Prior to the introduction of transgenic HT crop varieties, most growers used tillage to prepare the soil for planting. Excessive tillage, however, is known to cause soil structure changes, increase the susceptibility to soil erosion, and reduce soil moisture. Loss of topsoil due to tillage therefore causes environmental damage that can last for centuries. Since the early 1990s, growers have been reducing their tillage operations for soil conservation benefits. According to USDA survey data, about 60% of the area planted with GMHT soybean was under conservation tillage in 1997, compared with only about 40% for conventional soybean [170]. Gianessi [171] cites a survey by the American Soybean Association, indicating that U.S. soybean growers reported making fewer tillage passes through their fields since 1995 when GMHT soybean was first introduced. Because weed control can be done during the post-emergence phase, farmers can use direct-seeding techniques since there is no need for pre-seeding tillage. Conservation tillage leaves a layer of plant residues on the soil surface, preventing soil erosion, reducing evaporation and increasing the ability of the soil to absorb moisture [169]. A richer soil biota develops that can improve nutrient recycling and this may also help combat crop pests and diseases [142]. Earthworm populations are generally higher in no-till fields than in conventionally tilled fields [169]. In addition to a reduction in soil erosion and degradation, less frequent soil cultivation also results in a decrease in the emission of greenhouse gases, partly arising from a reduction in fuel use [154]. There is also evidence that conservation tillage can provide a wide range of benefits to farmland biodiversity by improving agricultural land as habitat for wildlife. The greater availability of crop residues and weed seeds can improve food supplies for insects, birds, and small mammals [142].

8
Scientific Debates on the Ecological Impact of GM Crops

The interpretation of collected scientific data is debated controversially by different stakeholders involved in the debate on potential impact of GM crops on biodiversity. Although some groups argue that experience and solid scientific knowledge are still lacking, the ongoing debate is generally not purely due to a lack of scientific data, but more to an ambiguous interpretation of what is considered an ecologically relevant effect of GM crops. The interpretation of study results is thereby often challenged by the absence of a defined baseline for the evaluation of environmental effects of GM crops. Consequently, some consider any effect related to GM crops as being undesired, while others compare it to effects caused by modern agricultural practices recognizing that a multitude of factors involved cause environmental effects. The interpretation of study results is further often challenged by knowledge gaps on the natural variation occurring in any biological system. Rather than the GM crop alone being the influencing factor, environmental effects are caused by agricultural production systems where the GM crop is one factor among others. Although science can help to assess these natural variations, it will most probably not be possible to elucidate all ecological interactions taking place in such systems. In practice, decision-making will thus have to be not purely based on scientific criteria, but will also be strongly influenced by political, social, economical and ethical factors. Ecologically significant effects are only judged unacceptable (i.e., representing a damage) by the society if they are perceived as being linked to a deterioration in quality of a particular entity (e.g., biodiversity).

Valuation of scientific data is thus influenced by the individual and subjective perceptions of the terms safety, risk and uncertainty by the society and particularly by the persons involved in decision-making. The following list intends to highlight a number of issues, which mainly in Europe are currently debated controversially in the discussion on the safety of GM crops.

Effects of GM Crops on Non-target Organisms

- There is scientific controversy on the baseline that should be applied when assessing potential effects of insect-resistant GM crops. It is discussed whether this should be the most common agricultural practice used (e.g., integrated pest management), a practice like organic farming, which is only practiced by a low number of farmers, or a (hypothetical) practice that may represent the optimal system for the environment.
- There is a debate to what extent indirect toxic effects, i.e., effects on natural enemies that largely depend on the target pest, should be valuated considering that such effects are common for all pest control methods and not restricted to the use of insect-resistant GM crops.

Impacts of GM Crops on Soil Ecosystems

- A commonly accepted definition for soil quality has not yet been found.
- Population sizes and community structure of soil microorganism are subject to high variation, and the baseline comparison for ecological implication is still not clear. Standard indicator species have not been defined. Different studies use a range of different parameters and techniques.
- Is the presence of low percentages of activated transgenic *Bt*-toxin(s) from *Bt*-crops in soils a reason for concern, considering that *Bt*-toxins are naturally occurring in soils due to the soil bacteria *Bacillus thuringiensis*, and that *Bt*-spray formulations are commonly used for insect control in agriculture and forestry?

Gene Flow from GM Crops to Wild Relatives

- In most agricultural landscapes, there is usually a gradual transition from peri-agricultural to semi-natural habitats. Although "wild plants" can usually be distinguished from "agricultural weeds", a clear definition of what plant species are considered being truly wild plants is lacking.
- Should effects occurring within agricultural or peri-agricultural environments be given the same importance as those effects, which could occur in natural habitats?
- Should gene flow from GM crops to wild relatives be valuated in a different way than gene flow from conventional crops to wild relatives?

Invasiveness of GM Crops into Natural Habitats

- Is the presence of volunteer GMHT oilseed rape in habitats such as field borders or road verges an unwanted environmental effect, considering that non-transgenic oilseed rape is regularly occurring in such habitats and that HT is not considered to confer a selective advantage in natural habitats?

Impacts of GM Crops on Pest and Weed Management and their Ecological Consequences

- Is it better to have a high biodiversity in-crop (i.e., to have weedy crops), or to enhance off-crop biodiversity (e.g., separate buffer strips outside the fields) providing food for insects and birds?
- Should herbicide-resistant weeds that have been caused by GMHT crops be valuated differently than herbicide-resistant weeds that have been caused by conventional (non-transgenic) weed management?

9 Conclusions

The risks of GM crops for the environment, and especially for biodiversity, have been extensively assessed worldwide over the past 10 years of commercial cultivation of GM crops. Consequently, substantial scientific data on environmental effects of the currently commercialized GM crops are available today, and will further be obtained given that several research programmes are underway in a number of countries. The data available so far provide no scientific evidence that the commercial cultivation of GM crops has caused environmental impacts beyond the impacts that have been caused by conventional agricultural management practices. Nevertheless, a number of issues related to the interpretation of scientific data on effects of GM crops on the environment are debated controversially. To a certain extent, this is due to the inherent fact that scientific data are always characterized by uncertainties, and that predictions on potential long-term or cumulative effects are difficult. Uncertainties can either be related to the circumstance that there is not yet a sufficient data basis provided for an assessment of consequences (the "unknown"), or to the fact that the questions to solve are out of reach for scientific methods (the "unknowable"). There is thus a need to develop scientific criteria for the evaluation of effects of GM crops on the environment in order to assist regulatory authorities when deciding whether environmental effects of GM crops are considered to represent a relevant environmental impact.

Agricultural production systems are complex and diverse. As with the adoption of any new technology, the use of agricultural biotechnology might include positive and possibly less favorable environmental impacts. GM cropping systems can help to reduce some environmental impacts associated with conventional agriculture, but they will also introduce new challenges that must be addressed. When discussing the risks of GM crops, one has to recognize that the real choice for farmers and consumers is not between a GM technology that may have risks and a completely safe alternative. The real choice is between GM crops and current conventional pest and weed management practices, all possibly having positive and negative outcomes. To ensure that a policy is truly precautionary, one should therefore compare the risk of adopting a technology against the risk of not adopting it [172]. We thus believe that both benefits and risks of GM crop systems should be compared with those of current agricultural practices.

Acknowledgements We would like to thank the Swiss Expert Committee for Biosafety for major funding of this review. We further thank Michèle Stark for help on an early draft of the manuscript.

References

1. Dale PJ, Clarke B, Fontes EMG (2002) Nat Biotechnol 20:567
2. Wolfenbarger LL, Phifer P (2000) Science 290:2088
3. Conner AJ, Glare TR, Nap J-P (2003) Plant J 33:19
4. Pretty J (2001) Environ Conserv 28:248
5. Snow AA, Andow DA, Gepts P, Hallerman EM, Power A, Tiedje JM, Wolfenbarger LL (2005) Ecol Appl 15:377
6. Stoate C, Boatman ND, Borralho RJ, Carvalho CR, de Snoo GR, Eden P (2001) J Environ Manage 63:337
7. Hails RS (2002) Nature 418:685
8. Robinson RA, Sutherland WJ (2002) J Appl Ecol 39:157
9. Chapin FS, Zavaleta ES, Eviner VT, Naylor RL, Vitousek PM, Reynolds HL, Hooper DU, Lavorel S, Sala OE, Hobbie SE, Mack MC, Diaz S (2000) Nature 405:234
10. Tilman D, Cassman KG, Matson PA, Naylor R, Polasky S (2002) Nature 418:671
11. Ammann K (2005) Trend Biotechnol 23:388
12. McLaughlin A, Mineau P (1995) Agric Ecosyst Environ 55:201
13. Walter T, Buholzer S, Kühne A, Schneider K (2005) In: Herzog F, Walter T (eds) Evaluation der Ökomassnahmen Bereich Biodiversität. Agroscope FAL Reckenholz. Eidgenössische Forschungsanstalt für Agrarökologie und Landbau, Zürich (Schriftenreihe der FAL Nr. 56)
14. Chamberlain DE, Fuller RJ, Bunce RGH, Duckworth JC, Shrubb M (2000) J Appl Ecol 37:771
15. Royal Society (2003) GM crops, modern agriculture and the environment. The Royal Society, London, p 17
16. Jaffe G (2004) Transgen Res 13:5
17. CFIA (2004) Canadian Food Inspection Agency, Ottawa, p 33
18. EFSA (2004) Guidance document of the scientific panel on genetically modified organisms for the risk assessment of genetically modified plants and derived food and feed. European Food Safety Authority, Brussels, p 94
19. EPA (1998) Guidelines for ecological risk assessment. US Environ Protection Agency, Washington, DC, p 80
20. OECD (1993) Safety considerations for biotechnology: scale-up of crop plants. Organisation for Economic Co-Operation and Development, Paris, p 43

21. James C (2005) Global status of commercialized biotech/GM crops: 2005, ISAAA Brief No. 34. International Service for the Acquisition of Agri-biotech Applications, Ithaca, NY, p 11
22. de Maagd RA (2004) In: Nap JPH, Atanassov A, Stiekema WJ (eds) Genomics for biosafety in plant biotechnology. IOS Press, Amsterdam, p 117
23. Ely S (1993) In: Entwistle PF, Cory JS, Bailey MJ, Higgs S (eds) *Bacillus thuringiensis*, an environmental biopesticide: theory and practice. Wiley, Chichester, p 105
24. Shelton AM, Zhao JZ, Roush RT (2002) Ann Rev Entomol 47:845
25. Pilcher CD, Rice ME, Obrycki JJ, Lewis LC (1997) J Econ Entomol 90:669
26. Gonzales-Nunez M, Ortego F, Castanera P (2000) J Econ Entomol 93:459
27. Dutton A, Romeis J, Bigler F (2005) Entomol Experiment Appl 114:161
28. AGBIOS (2006) Biotech Crop Database. AGBIOS, Merrickville, Ontario www.agbios.com
29. Ward DP, de Gooyer TA, Vaughn TT, Head G, McKee MJ, Astwood JD, Pershing JC (2005) In: Vidal S, Kuhlmann U, Edwards CR (eds) Western corn rootworm: ecology and management. CABI Publishing, Wallingford UK, p 239
30. Romeis J, Meissle M, Bigler F (2006) Nat Biotechnol 24:63
31. Hilbeck A, Baumgartner M, Fried PM, Bigler F (1998) Environ Entomol 27:480
32. Hilbeck A, Moar WJ, Pusztai-Carey M, Filippini A, Bigler F (1998) Environ Entomol 27:1255
33. Hilbeck A, Moar WJ, Pusztai-Carey M, Filippini A, Bigler F (1999) Entomol Experiment Appl 91:305
34. Dutton A, Klein H, Romeis J, Bigler F (2002) Ecol Entomol 27:441
35. Rodrigo-Simon A, De Maagd RA, Avilla C, Bakker PL, Molthoff J, Gonzalez-Zamora JE, Ferre J (2006) Appl Environ Microbiol 72:1595
36. Romeis J, Dutton A, Bigler F (2004) J Insect Physiol 50:175
37. Dutton A, Romeis J, Bigler F (2003) Biocontrol 48:611
38. Eizaguirre M, Albajes R, Lopez C, Eras J, Lumbierres B, Pons X (2006) Transgen Res 15:1
39. Pilcher CD, Rice ME, Obrycki JJ (2005) Environ Entomol 34:1302
40. Riddick EW, Dively G, Barbosa P (1998) Ann Entomol Soc USA 91:647
41. Daly T, Buntin GD (2005) Environ Entomol 34:1292
42. Naranjo SE (2005) Environ Entomol 34:1193
43. Whitehouse MEA, Wilson LJ, Fitt GP (2005) Environ Entomol 34:1224
44. Naranjo SE, Head G, Dively GP (2005) Environ Entomol 34:1178
45. Clark BW, Phillips TA, Coats JR (2005) J Agric Food Chem 53:4643
46. O'Callaghan M, Glare TR, Burgess EPJ, Malone LA (2005) Ann Rev Entomol 50:271
47. Candolfi MP, Brown K, Grimm C, Reber B, Schmidli H (2004) Biocontrol Sci Technol 14:129
48. Meissle M, Lang A (2005) Agric Ecosyst Environ 107:359
49. Musser FR, Shelton AM (2003) J Econ Entomol 96:71
50. Men XY, Ge F, Edwards CA, Yardim EN (2004) Phytoparasitica 32:246
51. Bambawale OM, Singh A, Sharma OP, Bhosle BB, Lavekar RC, Dhandapani A, Kanwar V, Tanwar RK, Rathod KS, Patange NR, Pawar VM (2004) Curr Sci 86:1628
52. Hagerty AM, Kilpatrick AL, Turnipseed SG, Sullivan MJ, Bridges WC (2005) Environ Entomol 34:105
53. Wu KM, Guo YY (2003) Environ Entomol 32:312
54. de la Poza M, Pons X, Farinos GP, Lopez C, Ortego F, Eizaguirre M, Castanera P, Albajes R (2005) Crop Protect 24:677

55. Head G, Moar M, Eubanks M, Freeman B, Ruberson J, Hagerty A, Turnipseed S (2005) Environ Entomol 34:1257
56. Torres JB, Ruberson JR (2005) Environ Entomol 34:1242
57. EPA (2001) Biopesticides registration action document – *Bacillus thuringiensis* plant-incorporated protectants. US Environmental Protection Agency, Washington DC, p 481
58. Babendreier D, Kalberer NM, Romeis J, Fluri P, Mulligan E, Bigler F (2005) Apidologie 36:585
59. Malone LA (2004) Bee World 85:29
60. Malone LA, Pham-Delegue MH (2001) Apidologie 32:287
61. Losey JE, Rayor LS, Carter ME (1999) Nature 399:214
62. Hellmich RL, Siegfried BD, Sears MK, Stanley-Horn DE, Daniels MJ, Mattila HR, Spencer T, Bidne KG, Lewis LC (2001) Proc Natl Acad Sci USA 98:11925
63. Stanley-Horn DE, Dively GP, Hellmich RL, Mattila HR, Sears MK, Rose R, Jesse LCH, Losey JE, Obrycki JJ, Lewis L (2001) Proc Natl Acad Sci USA 98:11931
64. Anderson PL, Hellmich RL, Sears MK, Sumerford DV, Lewis LC (2004) Environ Entomol 33:1109
65. Oberhauser KS, Prysby MD, Mattila HR, Stanley-Horn DE, Sears MK, Dively G, Olson E, Pleasants JM, Lam WKF, Hellmich RL (2001) Proc Natl Acad Sci USA 98: 11913
66. Sears MK, Hellmich RL, Stanley-Horn DE, Oberhauser KS, Pleasants JM, Mattila HR, Siegfried BD, Dively GP (2001) Proc Natl Acad Sci USA 98:11937
67. Dively GP, Rose R, Sears MK, Hellmich RL, Stanley-Horn DE, Calvin DD, Russo JM, Anderson PL (2004) Environ Entomol 33:1116
68. Baumgarte S, Tebbe CC (2005) Molec Ecol 14:2539
69. Saxena D, Flores S, Stotzky G (1999) Nature 402:480
70. Zwahlen C, Hilbeck A, Gugerli P, Nentwig W (2003) Molec Ecol 12:765
71. Mendelsohn M, Kough J, Vaituzis Z, Matthews K (2003) Nat Biotechnol 21:1003
72. Dubelman S, Ayden BR, Bader BM, Brown CR, Jiang CJ, Vlachos D (2005) Environ Entomol 34:915
73. Saxena D, Stotzky G (2000) FEMS Microbiol Ecol 33:35
74. Stotzky G (2004) Plant Soil 266:77
75. Tapp H, Stotzky G (1998) Soil Biol Biochem 30:471
76. Venkateswerlu G, Stotzky G (1992) Curr Microbiol 25:225
77. Head G, Surber JB, Watson JA, Martin JW, Duan JJ (2002) Environ Entomol 31:30
78. Hopkins DW, Gregorich EG (2003) Eur J Soil Sci 54:793
79. Herman RA, Evans SL, Shanahan DM, Mihaliak CA, Bormett GA, Young DL, Buehrer J (2001) Environ Entomol 30:642
80. Sims SR, Ream JE (1997) J Agric Food Chem 45:1502
81. Bruinsma M, Kowalchuk GA, van Veen JA (2003) Biol Fert Soils 37:329
82. Widmer F (2007) Assessing Effects of Transgenic Crops on Soie Microbial Communities (in this volume). Springer, Heidelberg
83. Cartwright C, Lilley A, Kirton J (2004) Mechanisms for investigating changes in soil ecology due to GMO releases. DEFRA Department for Environment Food and Rural Affairs, London, p 133
84. Motavalli PP, Kremer RJ, Fang M, Means NE (2004) J Environ Qual 33:816
85. Widmer F, Oberholzer HR (2003) In: Francaviglia R (ed) Agricultural impacts on soil erosion and soil biodiversity: developing indicators for policy analysis. OECD Organisation for Economic Co-operation and Development, Rome, Italy, p 551

86. Griffiths BS, Caul S, Thompson J, Birch ANE, Scrimgeour C, Andersen MN, Cortet J, Messean A, Sausse C, Lacroix B, Krogh PH (2005) Plant Soil 275:135
87. Manachini B, Lozzia GC (2002) Boll Zool Agr Bachic Ser II 34:85
88. Saxena D, Stotzky G (2001) Soil Biol Biochem 33:1225
89. Escher N, Käch B, Nentwig W (2000) Basic Appl Ecol 1:161
90. Pont B, Nentwig W (2005) Biocontr Sci Technol 15:341
91. Wandeler H, Bahylova J, Nentwig W (2002) Basic Appl Ecol 3:357
92. Sims SR, Martin JW (1997) Pedobiologia 41:412
93. Yu L, Berry RE, Croft BA (1997) J Econ Entomol 90:113
94. Zwahlen C, Hilbeck A, Howald R, Nentwig W (2003) Molec Ecol 12:1077
95. Vercesi ML, Krogh PH, Holmstrup M (2006) Appl Soil Ecol 32:180
96. Ellstrand NC (2003) Dangerous liaisons? When cultivated plants mate with their wild relatives. The John Hopkins University Press, Baltimore
97. de Nijs HCM, Bartsch D, Sweet JB (2004) Introgression from genetically modified plants into wild relatives. CABI Publishing, Wallingford UK
98. Hails RS, Morley K (2005) Trend Ecol Evolut 20:245
99. Stewart CN, Halfhill MD, Warwick SI (2003) Nat Rev Genet 4:806
100. Ellstrand NC, Prentice HC, Hancock JF (1999) Ann Rev Ecol Syst 30:539
101. GM Sci Review Panel (2003) GM science review: first report. Department of Trade and Industry, London, p 296
102. Jacot Y, Ammann K (1999) In: Ammann K, Jacot Y, Simonsen V, Kjellson G (eds) Methods for Risk Assessment of Transgenic Plants III. Ecological risks and prospects of transgenic plants, vol 3. Birkhäuser Verlag, Basel, p 99
103. Mason P, Braun L, Warwick SI, Zhu B, Stewart CN (2003) Environ Biosafe Res 2:263
104. Vacher C, Weis AE, Hermann D, Kossler T, Young C, Hochberg ME (2004) Theor Appl Genet 109:806
105. Daniels R, Boffey C, Mogg R, Bond J, Clarke R (2005) The potential for dispersal of herbicide tolerance genes from genetically modified, herbicide tolerant oilseed rape to wild relative. Centre for Ecology and Hydrology (CEH) Dorset, Dorchester UK, p 23
106. Halfhill MD, Millwood RJ, Raymer PL, Stewart CN (2002) Environ Biosafe Res 1:19
107. Halfhill MD, Zhu B, Warwick SI, Raymer PL, Millwood RJ, Weissinger AK, Stewart CN (2004) Environ Biosafe Res 3:73
108. Hansen LB, Siegismund HR, Jørgensen RB (2001) Genet Resour Crop Evolut 48:621
109. Hansen LB, Siegismund HR, Jørgensen RB (2003) Heredity 91:276
110. Warwick SI, Simard MJ, Legere A, Beckie HJ, Braun L, Zhu B, Mason P, Seguin-Swartz G, Stewart CN (2003) Theor Appl Genet 107:528
111. Moyes CL, Lilley JM, Casais CA, Cole SG, Haeger PD, Dale PJ (2002) Molec Ecol 11:103
112. Darmency H, Lefol E, Fleury A (1998) Molec Ecol 7:1467
113. Gueritaine G, Bazot S, Darmency H (2003) New Phytologist 158:561
114. Gueritaine G, Sester M, Eber F, Chevre AM, Darmency H (2002) Molec Ecol 11:1419
115. Reichman JR, Watrud LS, Lee EH, Burdick CA, Bollman MA, Storm MJ, King GA, Mallory-Smith C (2006) Molec Ecol 15:4243
116. Rufener A, Mazyad P, Ammann K (1999) In: Ammann K, Jacot Y, Simonsen V, Kjellson G (eds) Methods for risk assessment of transgenic plants III. Ecological risks and prospects of transgenic plants, vol 3. Birkhäuser Verlag, Basel, p 95
117. Elton CS (1958) The ecology of invasions by animals and plants. Methuen and Co Ltd., London
118. Levine JM, Vila M, D'Antonio CM, Dukes JS, Grigulis K, Lavorel S (2003) Proc Royal Soc Lond Ser B 270:775

119. D'Antonio C, Meyerson LA (2002) Restor Ecol 10:703
120. Sakai AK, Allendorf FW, Holt JS, Lodge DM, Molofsky J, With KA, Baughman S, Cabin RJ, Cohen JE, Ellstrand NC, McCauley DE, O'Neil P, Parker IM, Thompson JN, Weller SG (2001) Ann Rev Ecol Systemat 32:305
121. Levin SA (1988) Trend Biotechnol 6:S47
122. Beckie HJ, Seguin-Swartz G, Nair H, Warwick SI, Johnson E (2004) Weed Sci 52:152
123. Hall L, Topinka K, Huffman J, Davis L, Good A (2000) Weed Sci 48:688
124. Beckie HJ, Warwick SI, Nair H, Seguin-Swartz GS (2003) Ecol Appl 13:1276
125. Legere A (2005) Pest Manage Sci 61:292
126. Hall LM, Habibur Rahman M, Gulden RH, Thomas AG (2005) In: Gressel J (ed) Crop ferality and volunteerism. CRC Press, Boca Raton, FL, p 59
127. Crawley MJ, Hails RS, Rees M, Kohn D, Buxton J (1993) Nature 363:620
128. Crawley MJ, Brown SL, Hails RS, Kohn DD, Rees M (2001) Nature 409:682
129. Warwick S, Stewart CN (2005) In: Gressel J (ed) Crop ferality and volunteerism. CRC Press, Boca Raton, FL, p 9
130. Heard MS, Rothery P, Perry JN, Firbank LG (2005) Weed Res 45:331
131. Watkinson AR, Freckleton RP, Robinson RA, Sutherland WJ (2000) Science 289:1554
132. Dewar AM, May MJ, Woiwod IP, Haylock LA, Champion GT, Garner BH, Sands RJN, Qi AM, Pidgeon JD (2003) Proc Royal Soc Lond Ser B 270:335
133. Firbank LG, Forcella F (2000) Science 289:1481
134. May MJ, Champion GT, Dewar AM, Qi A, Pidgeon JD (2005) Proc Royal Soc Ser B 272:111
135. Squire GR, Brooks DR, Bohan DA, Champion GT, Daniels RE, Haughton AJ, Hawes C, Heard MS, Hill MO, May MJ, Osborne JL, Perry JN, Roy DB, Woiwod IP, Firbank LG (2003) Philos Trans Royal Soc Lond Ser B 358:1779
136. Hawes C, Haughton AJ, Osborne JL, Roy DB, Clark SJ, Perry JN, Rothery P, Bohan DA, Brooks DR, Champion GT, Dewar AM, Heard MS, Woiwod IP, Daniels RE, Young MW, Parish AM, Scott RJ, Firbank LG, Squire GR (2003) Philos Trans Royal Soc Lond Ser B 358:1899
137. Chassy B, Carter C, McGloughlin M, McHughen A, Parrott W, Preston C, Roush R, Shelton A, Strauss SH (2003) Nat Biotechnol 21:1429
138. Freckleton RP, Sutherland WJ, Watkinson AR (2003) Science 302:994
139. Firbank LG, Heard MS, Woiwod IP, Hawes C, Haughton AJ, Champion GT, Scott RJ, Hill MO, Dewar AM, Squire GR, May MJ, Brooks DR, Bohan DA, Daniels RE, Osborne JL, Roy DB, Black HIJ, Rothery P, Perry JN (2003) J Appl Ecol 40:2
140. Carpenter J, Felsot A, Goode T, Hammig M, Onstad D, Sankula S (2002) Comparative environmental impacts of biotechnology-derived and traditional soybean, corn, and cotton crops. Council for Agricultural Science and Technology, Ames, Iowa, p 200
141. Phipps RH, Park JR (2002) J Animal Feed Sci 11:1
142. Holland JM (2004) Agric Ecosyst Environ 103:1
143. Heap I (2006) The International Survey of Herbicide Resistant Weeds. www.weedscience.com
144. Owen MDK, Zelaya IA (2005) Pest Manage Sci 61:301
145. Bradshaw LD, Padgette SR, Kimball SL, Wells BH (1997) Weed Technol 11:189
146. CFIA (2003) Technical workshop on the management of herbicide tolerant crops. Canadian Food Inspection Agency, Ottawa, p 28
147. VanGessel MJ (2001) Weed Sci 49:703
148. EPA (2006) Office of Pesticide Programs. www.epa.gov/pesticides/
149. Health Canada (2006) Pest Management Regulatory Agency. www.pmra-arla.gc.ca/english/index-e.html

150. HRAC (2006) Guideline to the management of herbicide resistance. Herbicide Resistance Action Committee, Brussels, Belgium, www.plantprotection.org/hrac/
151. Benbrook C (2001) Pesticide Outlook 12:204
152. Benbrook C (2003) BioTech InfoNet, p 42
153. Brimner TA, Gallivan GJ, Stephenson GR (2005) Pest Manage Sci 61:47
154. Brookes G, Barfoot P (2005) AgBioForum 8:187
155. Fernandez-Cornejo J, Mc Bride WD (2002) Economic Research Service. United States Department of Agriculture, Washington, DC, p 61
156. Duke SO (2005) Pest Manage Sci 61:211
157. Qaim M, Traxler G (2005) Agr Econ 32:73
158. Trigo EJ, Cap EJ (2003) AgBioForum 6:87
159. FAO (2004) The state of food and agriculture – agricultural biotechnology, meeting the needs of the poor? Food and Agric Organization of the United Nations, Rome, p 208
160. Raney T (2006) Curr Opin Biotechnol 17:174
161. Fitt GP, Wakelyn PJ, Stewart JM, James C, Roupakias D, Hake K, Zafar Y, Pages J, Giband M (2004) Global status and impacts of biotech cotton: report of the second expert panel on biotechnology of cotton. International Cotton Advisory Committee, Washington, DC, p 65
162. Wu KM, Guo YY (2005) Ann Rev Entomol 50:31
163. Pray CE, Huang JK, Hu RF, Rozelle S (2002) Plant J 31:423
164. Hossain F, Pray CE, Lu Y, Huang J, Fan C, Hu R (2004) Int J Occup Environ Health 10:296
165. Huang JK, Hu RF, Rozelle S, Pray C (2005) Science 308:688
166. Bennett R, Morse S, Ismael Y (2003) 7th ICABR International Conference, Ravello, Italy, p 11
167. Burnside OC (1996) In: Duke SO (ed) Herbicide-resistant crops. CRC Lewis Publishers, Boca Raton, FL, p 391
168. CCOC (2001) Canola Council of Canada, Winnipeg, p 55
169. Fawcett R, Towery D (2002) Conservation Technology Information Center, West Lafayette, IN, p 20
170. Fernandez-Cornejo J, Caswell M (2006) United States Department of Agriculture–Economic Research Service, Washington, DC, p 30
171. Gianessi LP (2005) Pest Manage Sci 61:241
172. Goklany IM (2002) Nat Biotechnol 20:1075
173. Donegan KK, Palm CJ, Fieland VJ, Porteous LA, Ganio LM, Schaller DL, Bucao LQ, Seidler RJ (1995) Appl Soil Ecol 2:111
174. Palm CJ, Schaller DL, Donegan KK, Seidler RJ (1996) Can J Microbiol 42:1258
175. Sims SR, Holden LR (1996) Environ Entomol 25:659
176. Mikkelsen TR, Andersen B, Jørgensen RB (1996) Nature 380:31
177. Hauser TP, Shaw RG, Østergard H (1998) Heredity 81:429
178. Hauser TP, Jørgensen RB, Østergard H (1998) Heredity 81:436
179. Snow AA, Andersen B, Jørgensen RB (1999) Mol Ecol 8:605
180. Pertl M, Hauser TP, Damgaard C, Jørgensen RB (2002) Heredity 89:212
181. Hauser TP, Damgaard C, Jørgensen RB (2003) Am J Botany 90:571
182. Halfhill MD, Sutherland JP, Moon HS, Poppy GM, Warwick SI, Weissinger AK, Rufty TW, Raymer PL, Stewart CN (2005) Mol Ecol 14:3177
183. Ammitzbøll H, Mikkelsen TN, Jørgensen RB (2005) Environ Biosafe Res 4:3
184. Rieger MA, Potter TD, Preston C, Powles SB (2001) Theor Appl Genet 103:555

Invited by: Professor Sautter

Author Index Volumes 101–107

Author Index Volumes 1–50 see Volume 50
Author Index Volumes 51–100 see Volume 100

Subject Index

Printing: Krips bv, Meppel
Binding: Stürtz, Würzburg